THE L
SCIENCE OF MEASURING THE EARTH

ADVENTURES UNLIMITED PRESS

THE LOST SCIENCE OF MEASURING THE EARTH

Discovering the Sacred Geometry of the Ancients

Robin Heath
John Michell

The Lost Science of Measuring the Earth

Published in Britain as *The Measure of Albion*

Published by
Adventures Unlimited Press
Kempton, Illinois 60946 USA

www.adventuresunlimitedpress.com

ISBN 1-931882-50-9

Printed in the United States of America

10 9 8 7 6 5 4 3 2 1

THE LOST SCIENCE OF MEASURING THE EARTH

Robin Heath
John Michell

Acknowledgements

The authors would like to thank John Neal, Dr Euan MacKie, Richard Heath, Paul Broadhurst, John Martineau, Adrian Gilbert, Tricia Osborne and Sarah Sharp for their guidance and comments during the preparation and writing of this book. We would also like to thank Dr Raymond Garlick for his kind permission to use an extract from one of his poems at the beginning of chapter five. Other acknowledgements are given at their appropriate places in the text.

Illustrations

Most of the original graphical illustrations are by the authors. Unless otherwise indicated, photographs are by Robin Heath. The map on page 57 is taken from *Prehistoric Preseli,* by N P Figgis, and reproduced with her kind permission. The artwork below is of *Llech y Drybedd* dolman mirrored against *Carn Ingli,* its sacred summit. The artwork on page (v) is of *Pentre Ifan* also mirrored against *Carn Ingli*. These monuments are located in the Preseli mountains in Pembrokeshire, West Wales, and both artworks are by local artist Maura Hazelden, and are reproduced here with her kind permission.

Cover Illustrations

Front cover: Sunset in the Denbighshire mountains, Mold, North Wales.
Rear cover: Midwinter sunset at Stonehenge, taken from the Heel stone.

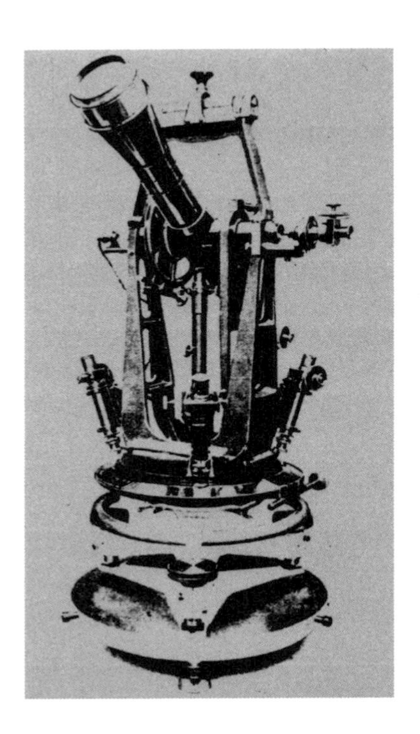

An Ordnance Survey Theodolite from 1885. A masterpiece of precision, this primary standard instrument was used for the surveys that produced the first editions of one inch to one mile OS maps, and is capable of measuring angles to fractions of a second* of degree. It was not the first time that these islands had been surveyed - Britain had been measured and surveyed to high accuracy over 5,000 years previously.

*One second of a degree represents a separation of about one foot viewed from 39 miles away.

THE MEASURE OF ALBION

CONTENTS

Foreword

by Paul Broadhurst

Stonehenge remains the most visited heritage site in Britain. Hosting nearly a million visitors a year, this noble anachronism from another time has held on to its secrets with a remarkable tenacity. The monument has revealed remarkably few true facts about itself or the culture that erected it.

The prehistory of Lundy Island is even more obscure. The place itself eludes our attention. A visitor to Devon or Cornwall, or the southern coast of Wales, may be fortunate enough to catch glimpses of the island hovering on the edge of the horizon, where sea meets sky. Sometimes this grey spectre fades and becomes invisible, vanishing completely from our vision like some Celtic Otherworld, its existence veiled from the mundane senses, its secrets safe from the modern world.

Who could ever have imagined that these two places, so very different in every respect, are linked by a vast geometric pattern spread out across Britain? Or that this pattern, part of a survey of Britain, was laid out in prehistoric times to reflect a cosmic rhythm that governs life on Earth, that of the relationship between Sun and Moon? It seems an extraordinary, even outrageous, claim. Yet in the following pages we are presented with tangible evidence that this is indeed the case, and that ancient people were the bearers of profound skills, knowledge and vision. Their vision is now laid out before us, and it challenges any lingering doubts we may have harboured about the sophistication of human culture thousands of years ago.

There is an important message threading through this remarkable book. The world around us is in an apparently hopeless state of imbalance, with all its traditional structures breaking down. Yet the striking discovery described in this book demonstrates that the ancients located themselves on the Earth under a harmonious order driven by higher Laws they understood better than we do today. Higher, certainly more holistic cultural priorities can and once did occupy human endeavour. The pinnacle of human achievement is not necessarily now; a humbling realisation, and a lesson from which we must learn.

If we are ever to reconnect with the true mysteries of Creation and the manner in which they manifest, an essential start must be to rediscover the knowledge that guided our ancestors. We must acknowledge that in the twilight of the past much was known that has now been forgotten. We have lost touch with our roots. Humankind urgently needs to re-establish cultures with reference to higher, timeless principles. We need a plan to re-align with Creation.

In *The Measure of Albion,* two of the most experienced and respected researchers into the wisdom of the ancients each present their own perspective on a discovery, one which reveals how ancient wisdom was preserved in code within the landscape of Old Albion. It is almost as if it was intended to reappear to us at a time when most needed, to remind us of these higher principles that once guided our ancestors. Perhaps generations to come will again aspire to their vision; to live in harmony with Natural Law.

Figure 1.1 Silbury Hill, viewed from West Kennet long barrow. This great artificial mound, built of chalk blocks, is a monumental relic of the prehistoric surveyors who located the sacred places of Britain and unified them by their mysterious science of number, music and geometry.

The Discovery

A personal introduction by John Michell

One thing leads to another - once you have made a start. My own progress into the mysteries of ancient Britain began in the 1960s with the anomalous UFO phenomenon. Strange things were happening, or being said to happen: sightings of unknown lights and globes in the sky, and encounters with unearthly creatures. These things were real - in their effects on minds and popular culture. But the authorities were scornful. Scientists and government spokesmen denied the existence of UFOs and implied that anyone who experienced or took interest in the phenomenon was deluded, deceitful or feeble-minded.

What did they know that made them so sure? I wondered. What were they hiding from us? Then I grew older and met some of these authorities, and I saw that they were hiding nothing because they knew nothing. They were surprised that anyone should even mention the subject of UFOs.

This impelled me in 1967 to write *The Flying Saucer Vision*, about UFOs as portents of changing times and impending revelations. That is how C.G. Jung, in his remarkably prophetic UFO book of 1959 (*Flying Saucers: A Modern Myth of Things Seen in the Skies*), interpreted the phenomenon, and he has since been proved right. Nothing has been discovered about UFOs on the physical level, but as Jung saw them, as catalysts of a radical change in human thinking, they have been crucially effective. From UFO studies came a renewed interest in mysteries and anomalies generally. My own direction was into prehistoric archaeology - stone circles, megalithic sites and their alignments. Evidence of a lost science behind the siting of these relics led to studies of traditional geomancy, units of measure and number symbolism, and finally to the perception of an ancient, forgotten code of knowledge that, at certain times, has prevailed in countries all over the world.

Plato and other initiated writers allude to a strange, mystical science, known at different times and places in the past, but no longer openly practised. It was based on a code of knowledge, reputedly of divine origin, and wherever it appeared it attracted the blessings of heaven upon earth. Its secrets were long preserved by the priests of old Egypt. It was, they said, the instrument by which they had maintained their civilization, unchanged and with the same high cultural standards, for many thousands of years.

The key to this science revealed itself in 1970. I had been studying ancient, symbolic numbers, especially those that occur repeatedly in myths and allegories as well as in music, architecture, astronomy and the gematria of gnostic scriptures. Certain numbers kept appearing in different contexts. They were geometrically related, like components of the same inclusive diagram. Finally the diagram itself appeared. It turned out to be a cosmological plan showing, among many other things, the rational squaring of the circle, the cycle of the astronomers' *great year* and the earth and moon in their correct dimensions. The immediate clue behind this realization was St John's account of the New Jerusalem in Revelation 21, complete with its symbolic measures. The revealed cosmological plan was therefore called the Heavenly City diagram.

It later appeared that this diagram was the core of a greater figure which represents the traditional world-image, a rationalized, numerically-framed depiction of the cosmos as a divine, living creature. From this diagram came the laws and social structure of those nations which, at different times in history, have received that culture-bearing revelation and have adopted it as their ruling standard. Today again it is active among us, provoking a timely reaction against the modern, secular world-view and the chaos that emanates from it.

One practical benefit that has been derived from the holy city diagram is the discovery in it of a unified number code behind all the units of measure in the ancient world, and the exact definitions of these units. This made possible the precise measuring of Stonehenge and the recovery of the earth's dimensions, as formerly established. The next stage was to apply these measures - all based on the standard foot or mile - to the pattern of sacred sites around Stonehenge, and then further across the country. These local discoveries developed into the greater realization which is the main subject of this book.

My own approach to this realization has been as follows. For a long time I had been interested in the large-scale works of prehistoric surveyors across the British Isles, some of them practical land divisions and others apparently symbolic or magical. Sometimes, perhaps always, the practical and the magical coincide. A clear example is the arrangement of the four provinces of

Ireland, whose original sea borders were so placed that, when joined through the geographical centre, they create the twelve-part, astrological figure that symbolized the sacred constitution of Old Ireland.

A further feature of ancient surveying is the drawing of axis-lines through whole countries, symbolizing the divinely revealed standard of law that is the axis-pole of all cosmologically ordered societies. Best known of these is the St Michael axis, the original line of Icknield Street, linking the eastern and western extremities of southern England through its centre at Avebury.

Another example is the circle of Perpetual Choirs, a ten-sided figure, with nodes at Stonehenge, Glastonbury Abbey and other remarkable sites, whose centre is at a lonely spot in the Malvern Hills, called Whiteleafed Oak, where the counties of Herefordshire, Worcester and Gloucester have their meeting-point. One of the ten sides within of the Perpetual Choirs circle is formed by the direct line between Stonehenge and the site of the Celtic shrine at Glastonbury, to the west. At right-angles to this line is the direct line from Stonehenge to Avebury, measuring one quarter of a degree of latitude or 17.28 miles.

Those facts were in my mind when, on a train back home from Stroud, where there had been a formative meeting of the Ley Hunter Society, I sat with Adrian Hyde, a researcher in leys and prehistoric landscapes who had been at the meeting. He showed me his maps and notes on that mysterious feature, the so-called Roman road that runs in a dead-straight line 26 miles from Bath eastwards, and is directed upon Silbury Hill. It is not just a road but defines ancient boundary lines and a section of the Wansdyke. It runs parallel to the Stonehenge-Glastonbury line and at right angles to the Stonehenge-Avebury axis.

In Chapter Ten is further reference to the Perpetual Choirs circle and of the wonderful scheme of measure, both practical and symbolic, in the degree of latitude that includes both Stonehenge and Avebury. I was gripped by the beauty and subtlety of this scheme, and in a state of euphoric obsession, when I was visited by Robin Heath and his publisher, our old friend John Martineau. Other friends turned up at the same time, including John Neal, who has continued and greatly expanded my work on ancient metrology. It was a convivial evening, and I had no chance to talk to Robin Heath until he was about to go.

I wanted to ask him about the long-distance line that he had discovered and written about, the line running due west from Stonehenge to Lundy Island in the Bristol Channel which lies at the same latitude. He had drawn another line at right angles to it, from Lundy to a point near the Preseli mountains where the bluestone rocks at Stonehenge were quarried. A line from that point to Stonehenge completed a Pythagorean triangle with sides in the proportion 5, 12, 13.

I was struck by the grand scale of this concept, and admired it as a work of imagination. But I could see no compelling reason to accept Heath's 'lunation' triangle, described in full in chapter four, as a genuine ancient artifact. If the ancient surveyors had deliberately planned it, they would have employed a standard unit of measure, and I did not then have the data needed to research that. So that evening, just as Robin was leaving, I asked him, What is the length of your Stonehenge-Lundy baseline? He gave me a close enough estimate, 123.4 miles, and I recognized that as a distance I had already seen in my Stonehenge-Avebury researches, The precise measure is 123.42857, more elegantly rendered as 864/7 miles. That identifies the common unit in Heath's lunation triangle as 36/7 miles. In terms of this unit, the sides of the lunation triangle are 10, 24 and 26. It is the same triangle as in the Stonehenge station rectangle, but 2,500 times larger.

This realization broadened the scale of my researches, and of Heath's also. Coincidentally, starting from different points, we had been investigating the same pattern, laid out across the surface of Britain by surveyors of a very distant age. It seemed appropriate to share our findings and to pursue the inquiry further in partnership. It seemed to us also that this discovery of a refined and accurate pattern, linking the extremities and off-shore islets of Britain within a regular geometric framework, is of such profound interest and importance, and yet so radically at variance with the current archaeological view of prehistoric culture and capabilities, that it needed to be presented in the most careful and accurate way possible. Every statement of fact in this book has been checked and, as far as possible, verified by both of us. Any conclusions we draw are, of course, open to dispute.

When your mental energies are fired up by a subject you attract, by some natural though inexplicable cause, insights, information, people and experiences relevant to your obsession. Soon after my evening with Robin, while working with total attention, every available hour of day and night, on the geometric survey of prehistoric Britain, I heard from Adrian Gilbert, the author of books on the mystery of past civilizations. He sent me an advance copy of his latest, entitled *The New Jerusalem,* with a note referring me to the brief appendix on its last pages. 'This will interest you', he wrote. And so it did.

Gilbert's appendix is an analysis of the shape and dimensions of Britain, as given by Julius Caesar in his account of the Gallic war. There is no clue to where Caesar found his information. He never went much farther into Britain than the Kent coast, so he must have been quoting some forgotten authority, Greek, Roman or Celtic. But wherever they came from, the dimensions he attributes to the island of Britain, and the geometric shape that develops from them, are remarkably significant. Together with Robin Heath's inspired recognition of the line due north from Lundy island (which proves to be the main axis of the scheme), Caesar's figures have made it possible to reconstruct the great, mystical survey of prehistoric Britain.

On hearing of The Way,
the best of men explore its length

Lao Tzu

Figure 1.2 Stane Street, a prehistoric trackway, one of dozens of surviving aligned roads in Britain. Many such roads were subsequently adapted and widened for use by the Celts and the Romans.

THE MEASURE OF ALBION
- BOOK ONE -
Robin Heath

Chapter One

The Art of Surveying the Landscape

'Every journey conceals another journey within its lines:
the path not taken and the forgotten angle.'

Jeanette Winterson

This book investigates a system of land surveying (geodesy), practised throughout Britain in prehistoric times. Over the past three centuries this ancient science has been progressively ignored, the ancient traditions seen as worthless. One outcome of this has been the widely held view that the culture which once built the ancient trackways, standing stones and stone circles has nothing useful to tell us, it is merely for the history books.

A consequence of this change is that there is currently *no* explanation for the choice of sitings of ancient and prehistoric monuments. Lacking any knowledge of the *system* of ancient geodesy it has been all too easy to assume that the designs or their sitings were chosen arbitrarily and without prior thought or wisdom. Fortunately, many megalithic monuments have survived the deliberate and systematic destruction wrought by the various cultural and religious overlays imposed on these islands since their erection, and these monuments preserve both their design and their location within the landscape.

These remnants of prehistoric achievement in geodesy mark the peaks of a huge mountain of ancient geodetic science which had at its core an objective numerical system into which were incorporated clearly defined measures. Through the analysis of the dimensions and locations of surviving ancient sites, the authors have recovered both the system and the measures. The evidence, abundant and irrefutable, demonstrates that surveying and navigation skills were well developed and being applied in prehistoric Britain.

The more subjective aspects of ancient surveying, often called geomancy, can at best only support a *belief* in an underpinning system, they do not reveal that system. Yet despite the apparent absence of objectivity, geomancy has also survived, a pale and wan remnant of its former status. It has endured as folklore, which is usually considered quaint, often charming, and always capable of being enchanting. Recently, as books and films about magic, earth mysteries and ancient wisdom it has enjoyed an unparalleled revival of popularity.

Folklore is fed from a much older source than history. Many legends persist about stone circles, dragons on hill-tops, tunnels connecting ancient castles separated by unfeasible distances, and churches that would not allow themselves to be built except on a particular site. The oral traditions which have nourished this material down through the ages still nourish us today, while appearing stripped of any objective content. To many it is superstition, unscientific.

The sacred geometry to be found within old churches, stone circles, alignments and trackways provide another kind of evidence for the importance of land surveying, geometry and astronomy within both the ecclesiatistical and secular traditions since prehistoric times. It is rational and objective, *scientific*. The traditions of the Templars and of freemasonry have preserved some of this material, but incompletely, so that the underlying system remains obscure.

The Orthodox View of Ancestral Achievement

Until quite recently, the prehistoric inhabitants of Britain were deemed barbarians or savages. Because traditional archaeological evidence has not revealed it, the assumption has long been made that prehistoric Britons possessed no advanced numerical, geometrical or astronomical skills. Even when compelling evidence for these skills has been presented, the required changes to our present model of prehistory have been a 'path not taken' for historians.

Prehistoric astronomical alignments are presently admissable as *symbolic*, whilst the evidence for a precision astronomy is refuted. The shape of many stone rings is described as being 'ovoid', despite their exquisite and consistent geometry. The unit of length discovered at many megalithic sites by Professor Alexander Thom has largely been ignored, despite statistical analysis and confirmation of its length by renowned mathematicians.

The geodetic and geomantic evidence presented here indicates further levels of skill and aspiration previously unassociated with Neolithic, Bronze Age or even the later Celtic culture. These skills are, in essence, identical to those of modern surveying - the accurate measurement of the distance between two locations and the bearing, or angle, from the first location to the second.

Modern Geodesy

A modern navigator, surveyor or map-reader is primarily interested in measuring the distance apart and the angular bearing from point *a* to point *b* with reference to true north. This angle is called the *azimuth*. For distances exceeding a few miles, *the calculations require an accurate knowledge of the size and shape of the earth,* our modern science of geodesy. Just three centuries old this nascent science abandoned as inaccurate and unreliable a much older science, which we shall show to have been applied at least fifty centuries ago, and possibly further back still.

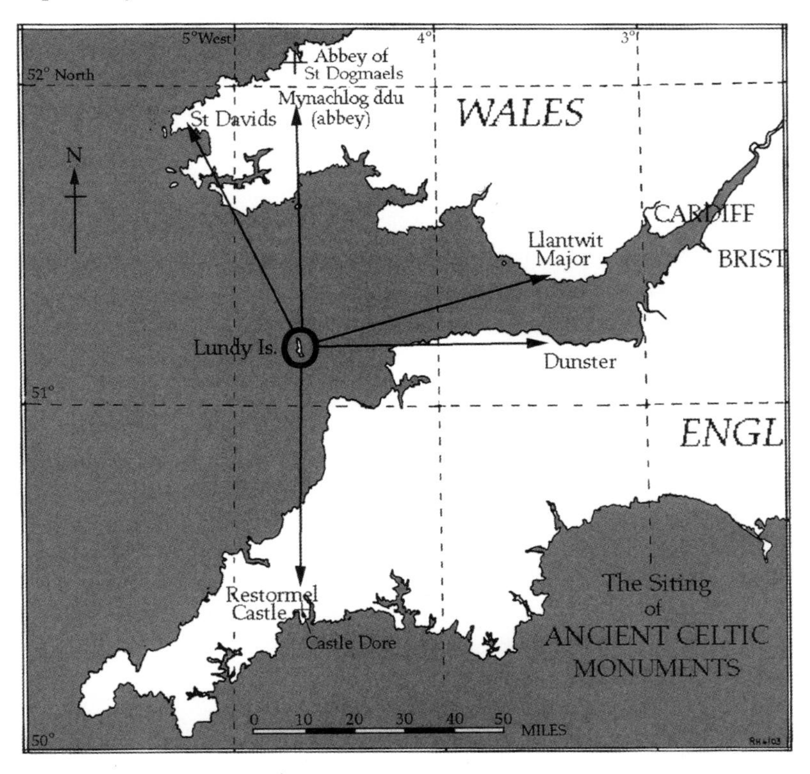

Figure 1.3 The central position of Lundy island to many Celtic monastic sites suggests a geodetic importance for both the island and the sites in question. The sites marked with arrows all lie within three per cent of the same distance from Lundy, an average distance of 54 miles. Four of the named sites above align to the cardinal directions of the compass and all sites supported Celtic monasteries and teaching centres. These sites are also inextricably bound up with the legend of the fifth century King Arthur and are sited adjacent to significant prehistoric monuments named after him.

(Adapted from an original illustration in *Sun, Moon & Stonehenge*.)

Isaac Newton's quest to determine accurately the size of the earth involved a study of ancient measures which tradition implied were related to the polar radius. Inadequate knowledge of this system of measures led Newton, in 1668, to an inaccurate result, ancient geodesy subsequently being pushed aside in a modern race to establish accurate figures for the size and shape of the earth. One outcome of this quest was the development of better instrumentation, another the improvements in the techniques of surveying which evolved during the nineteenth and twentieth centuries. Yet despite both these innovations, the values now acclaimed as being the definitive dimensions of the earth are discovered to be no more accurate than those employed by the ancients.

So, here's the rub. The ancients had somehow acquired accurate figures, fully listed in Chapter Two, for the key dimensions of the earth. From where did they get this information - did they measure the planet themselves or did they inherit these astonishingly accurate figures from an even earlier prehistoric culture? In our search for answers to these questions the authors have been able to demonstrate that the measures employed by the ancient world relate to the size of the earth.

We made one initial assumption: that the sites remain in their original locations. This is possibly the sole point concerning ancient geodesy upon which historians, scientists, archaeologists, geomancers and ley-hunters alike would currently all agree!

The Work in Hand

The path we have taken, the 'forgotten angle', has been discovery led and tradition fed. To demonstrate that geodetic surveying to high accuracy was being practiced in prehistoric times, we adopted the following methods.

Firstly, significant ancient sites were located and surveyed, and the unit(s) of measure used within their construction identified, as known historical measures.

Secondly, the distances between the sites were measured accurately in order to identify whether the same or related units were being used to space the sites across the landscape.

Thirdly, the locations of sacred and ancient sites were studied to see if they revealed a definite message or meaning encoded by the ancients and decipherable by modern man.

As this exploration advanced, the accuracy of the original measurements indicated the precision and even suggested the methods by which the distances and angles between sites were established. One significance of such a discovery enlarges the present belief that prehistoric man revered the landscape to a new perspective which must now include the measurement of the earth's size and shape. A practical benefit of this research it that it can and does indicate the likely positions of significant archaeological sites to precision accuracy.

In addition, it has been possible to suggest a credible source for the length and reasons for use of the so-called megalithic yard, that much disputed unit of length discovered by Alexander Thom and first revealed in his *Megalithic Sites in Britain* (Oxford, 1967). This unit will be shown to integrate perfectly within the canon of ancient metrology, while being objectively derived from astronomical observation. The key time cycles affecting life on earth will be seen to have been incorporated into the known lengths of antiquity.

We first turn our attention to demonstrating that the ancients indeed understood that the earth was a spheroid and knew its key dimensions. The proof may be found in ancient and not-so-ancient texts, and is encoded within the various units of length employed throughout the ancient world, and incorporated within their most significant surviving monuments.

Figure 1.4 Between 3000 and 1500 BC, thousands of stone circles were erected throughout northwestern Europe, yet their purpose remains a mystery. In this book, the dimensions, locations and distances separating prehistoric sites were accurately measured to shed light on why they were built and why they were so important to the Neolithic and Bronze Age cultures that built them. The illustration above shows *Boscawen-un* stone circle, Cornwall. Nineteen regularly spaced stones enclose a central pointed megalith, now leaning over perilously. It forms part of a long alignment through the west of the county, one of many throughout Britain.

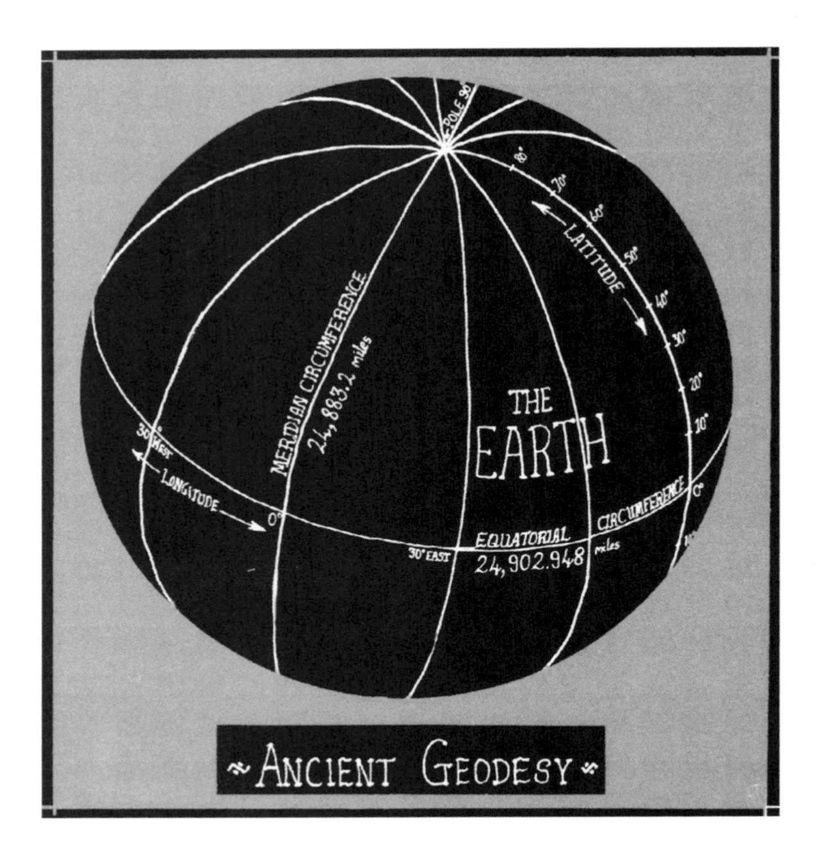

ANCIENT GEODETIC VALUES

	Radius	Circumference
POLAR	3,949.7142..	24,883.2 ($= \frac{12^5}{10}$)
MEAN	3,958.6909..	(MERIDIAN)
EQUATORIAL	3,963.42857..	24,902.94857..

Figure 2.1 The principal dimensions of the Earth according to ancient geodesy. The illustration shows (*top*) the meridian circumference and equatorial circumference together with meridian lines (equal longitudes). The dimensions in the table above are given in miles.

Opposite top - One of the Arundel Marbles in the Ashmolean Museum The Roman foot marked above the right arm is the *Statilian foot,* measured by John Greaves in 1639 as being 0.972 English feet. A century later, Matthew Raper obtained a value of 0.973 feet from the measurement of many Roman public buildings. The diameter of Stonehenge is 100 such feet.

Chapter Two

Measuring the World

'They discuss and teach youths about the heavenly bodies and their motions, the dimensions of the world and of countries, natural science and the powers of the immortal gods.'

Caesar, on the British Druids

Did prehistoric people measure length or distances accurately, and, if they did, how did they do it? Where are their rods or ropes and how did they maintain a standard of length? How could an agrarian tribal society that ate from crude pots become so motivated as to accurately measure angles or long lengths over often rugged terrain? Even more pertinent would be to enquire as to how was it possible to determine the size of the earth - could prehistoric surveyors have measured latitude and longitude?

The presently adopted model of prehistoric culture in Britain places an apparently insurmountable obstacle in answering such questions. Yet based on the evidence which we shall present, it becomes clear that prehistoric Britons had not only addressed the problems of surveying but had also solved them satisfactorily. However, before we can fully appreciate this evidence, we must first look at how our own epoch, the historical period from 2000 BC to modern times, unravelled the science of surveying and geodetics - how did *our* culture learn to measure the Earth and make maps?

The Classical Roots of Modern Geography

Whilst considerable information concerning Babylonian and Egyptian surveying techniques has been recovered, in the orthodox historical record it is a Greek, Anaximander, who is credited with the distinction of being the inventor of maps. Working during the 6th century BC, he is also thought to have been the first of many Greek philosophers to attempt the determination of the size of the earth. Noting that the whole sky revolved around the celestial pole, he declared that the earth was at the centre of a colossal sphere whose inside surface held the stars in place. Before a further century had passed, Thales of Miletus announced that the earth itself was spherical.

During this same period, Pythagoras also taught that the earth was a sphere. Why might he have thought this? The two most evident indications are firstly, that an observer on the shore sees a departing ship vanishing over the horizon, hull first followed by its sails and masts, and secondly, a lunar eclipse shows the shadow of the curved surface of the earth projected onto the moon. Such observations could have indicated that the earth's surface was curved, and, like the moon, spherical in shape.

It is also recorded that Pythagoras suggested that the earth orbited the sun, and rotated about its own axis, a troublesome fact that somewhat deflates our own more recent *hubris* in claiming to have discovered heliocentrism. Copernicus may have postulated, and Galileo subsequently proved the heliocentric facts of earthly life, yet Pythagoras and others preceded the pair of them by two millenia, literally confirming that there is nothing new under the sun.

Modern opinion suggests that the ideas of most philosophers who lived during the era of Pythagoras, Socrates, Aristotle and Plato, were based not on facts gained from observation or experiment, i.e. science, but on their philosophical ideas. For example, Pythagoras and his followers thought that the earth and the whole universe were built according to a perfect geometrical plan - a belief which is anathema to modern cosmologists. *Yet emerging from these Greek works come figures for the size of the Earth which are so astonishingly accurate that they suggest that their source derived from precision measurements.*

Pythagoras performed many scientific experiments, including the demonstration that the musical scale was comprised of notes whose frequencies of vibration each bear a simple numerical ratio to other notes of the scale. And it remains true that *all* practical trigonometry and surveying depends wholly on that famous theorem concerning right-angled triangles named after him.

The Size and Shape of the Earth

There are many ancient estimates of the size of the Earth, but they are mostly given in units of measure whose names and lengths remain unfamiliar to the modern world. We have forgotten the cubits, royal cubits, sacred cubits, remens and stades of the ancient world. Ancient metrology appears as a mad jumble of quaint units of length until one discerns the underpinning system.

Through applying the numerical methods employed by the Platonic school, John Michell was able to re-discover vitally important aspects of the system of ancient metrology. Using Michell's book *Ancient Metrology* as his base, John Neal then revealed the bigger picture found in *All Done With Mirrors.*

To understand this system at its most basic level, the reader has only to understand two simple truths:-

> *1. The ancient units of length are grouped in 'families', connected to each other by whole number ratios, based on a root measure whose length is always related to the English foot. The example below shows these interconnections, and the fractional ratios, which connect the foot to the various cubits of the ancient world.*
>
> *2. These 'families' of units of length form exact fractional parts to one of the principal dimensions of the Earth: polar radius, meridional circumference or equatorial circumference. In other words, they are commensurate with the dimensions of the earth.*

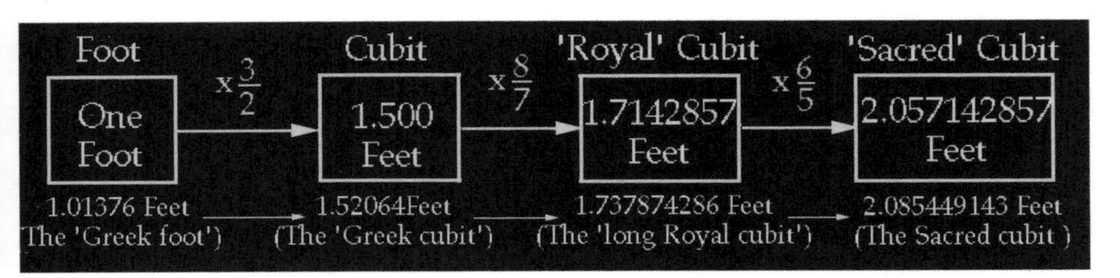

Figure 2.2 The route from the root foot to the sacred cubit, via the common cubit and royal cubit, showing the fractional conversion ratios. If Newton had used the sacred cubit derived from the 'Greek' foot of 1.01376 English feet in his calculations, a length of 2.085449143 feet (shown in the lower part of the diagram), and multiplied that by ten million, he would have found the correct size for the polar radius of the earth. Newton's inability to understand the system of ancient metrology led to the modern quest to discover the size and form of the earth, a quest which began from scratch and, after three centuries, has admirably demonstrated the astonishing accuracy of the ancient figures.

A simple example demonstrates the rational basis of this system. In the 2nd century AD, the great astronomer and astrologer Claudius Ptolemy wrote his epic *Geographia,* a work in eight parts. This treatise details a mathematical geography and the problems connected with projecting a plane surface, a map, onto a spherical earth. Ptolemy gives a figure for a one degree arc of the earth's surface as being 300,000 Roman remens in length (of 1.216512 English feet). The Roman 'geographic' remen is related to the Greek foot (1.01376 English feet) by the simple ratio 6:5. In feet, Ptolemy's degree of arc is therefore 364,953.6, which is 69.12 miles, the *present* figure for the meridian circumference degree and from which the Admiralty nautical mile is derived.

The Polar Radius

(Modern figure 3950.076 miles; Ancient figure 3949 $^{5}/_{7}$ miles)

Another example. Pliny recorded that the polar radius was 42,000 stades. This again means nothing today unless we know the length of the stade, which is 500 'polar' feet, each of 0.99307102 English feet. The polar foot is also known as the common Egyptian foot, and may be found at Stonehenge, where an exact 98 of such units define the inside diameter of the sarsen circle, as measured by Flinders Petrie at 97.325 English feet. The exact figure should be 97.32096 feet, just fifty-thousandths of an inch less.

Pliny's polar radius thus becomes a highly interesting 21,000,000 'polar' feet, an exact 3949 $^{5}/_{7}$ miles, precisely the ancient geodetic value, and a figure all but identical with the modern estimate.

The Meridian Circumference

(Modern and Ancient figure 24,883.2 miles)

To obtain the meridional circumference we cannot multiply the polar radius by twice pi, as we are not dealing with a true sphere. The earth is an ellipsoid and bulges at the equator, making the circumference through the poles, known as the meridian circumference, somewhat larger than it would be for a perfect sphere. We can show that the ancients applied this knowledge.

In *Ancient Metrology,* John Michell identified 24,883.2 miles as being the ancient geodetic value for the meridian circumference. To return to Ptolemy's figures, one degree of arc of latitude is 69.12 miles, and thus one minute of arc averages at 1.152 miles, or 6,082.56 feet, a value enshrined in the present day British nautical mile (and the knot), and which equals 6,000 Greek feet, affirming its historical geodetic origin. Multiplying this minute length by 21,600 (360 x 60) restores the meridian circumference of 24,883.2 miles.

The conversion between the ancient values of polar radius and meridian circumference is facilitated through the fraction 63/10, which effectively approximates 'pi' to 3.15. Such a conversion informs us not only that the polar radius and meridian circumference were known accurately in antiquity, but so too the shape of the earth.

The illustration overleaf confirms that 24,883.2 miles is exactly tailored to fit in with many ancient measures, this being the tip of a geodetic iceberg, for it is also mysteriously integrated with Imperial measures.

$$24{,}883.2 \text{ miles} = 12^5/10 \text{ English miles}$$

$$= 12^6 \times 44 \text{ English feet} = 131{,}383{,}296 \text{ feet}$$

We are no wiser as to how or from where these Greek philosopher-scientists acquired their figures, but now there remains no question that acquire them they certainly did.

Rationalising Pi

While Egyptian, Babylonian and Greek techniques are largely unfamiliar, written records indicate that Roman surveyors were using a highly developed technology to measure and map their expanding Empire. The methods and instruments by which they 'centuriated' all their territories are well attested. Significantly, the Romans and the earlier Greeks, Babylonians and Egyptians were using fractions to approximate, i.e. make rational, the irrational *pi*, in order to match perimeters to circumferences using whole number ratios. Cuneiform calculation tables for the volumes of circular wooden logs still exist, and they use $^{25}/_{8}$ for *pi*. Two millenia later, Vitruvius, in describing a Roman odometer writes,

'The wheels of the carriage are to be four feet in diameter and on one wheel a point is to be marked. When the wheel begins to move forward from this point and revolve on the road surface it will have completed a distance of 12 and a half feet on arriving at the point from which it began its revolution.'

This device was designed with a mechanism which audibly dropped a stone into a box slung under the wagon following the completion of every mile, after four hundred revolutions of the wheel, or one Roman mile. Because $^{25}/_{8}$ was used as the value for *pi*, the odometer would be 28 feet in error every 5,000 foot Roman mile, an unacceptable error. To correct this, the diameter was measured using the 'shorter' Roman foot of 0.96768 feet and the distance rolled out by the perimeter of the revolving wheel measured using the 'longer' Roman foot of 0.9732096 feet.

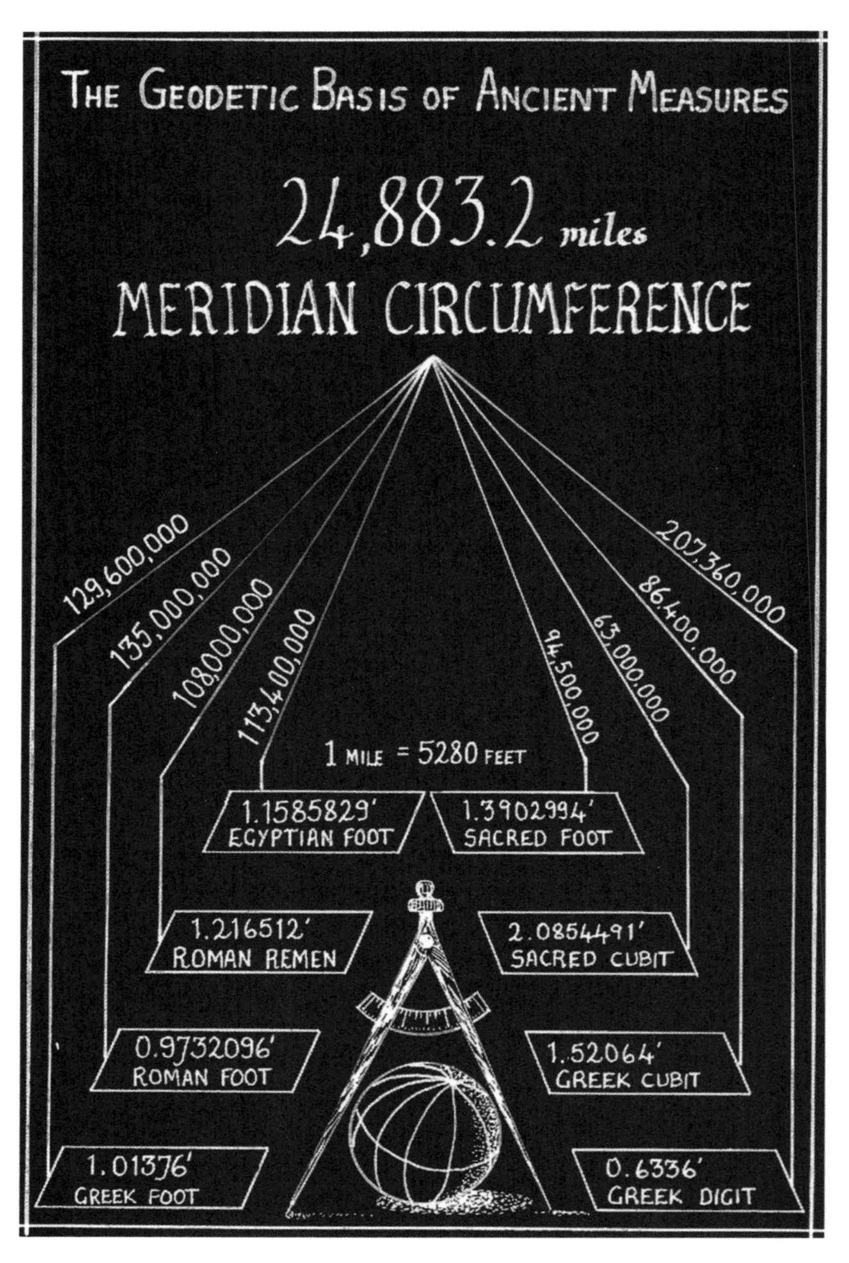

Figure 2.3 The geodetic basis for ancient measures. The measures of antiquity are grouped into families, many related to the meridian circumference of the Earth by exact whole number multipliers, some of which are illustrated above. This confirms their intended dimensions. 24,883.2 miles is also integrated with Imperial measures, being $12^5/10$ miles or $12^6 \times 44$ feet.

This neat trick to partly tame the irrational *pi* enabled the accurate measurement of distance using the odometer. Errors were minimised by using different lengths for the foot, a 'diameter' value and a 'perimeter' value, separated by the fraction $^{176}/_{175}$, a vital clue in understanding the system of ancient metrology, and one to which we will return.

In the historical record, we frequently find *pi* approximated to $^{25}/_{8}$, $^{25}/_{8}$ and larger fractions such as $^{864}/_{275}$ or $^{22,695}/_{85^2}$. In ancient geodesy, $^{63}/_{10}$ (two times 3.15) was used to obtain the meridional circumference from the polar radius (*see table below*).

'Pi' fraction	Fraction	Decimal Equivalent
Septenary	22/7	3.142857...
Octagonal	25/8	3.125
Polar radius to Meridional circumference conversion	63/10	3.15 x 2
True *pi*	N/A	3.14159265........

$^{22}/_{7} \times {}^{8}/_{25} = {}^{176}/_{175}$

The ancient values for the polar radius and meridional circumference were still known and in use by medieval navigators, as was the technique of using *pi* equal to $^{63}/_{20}$. Towards the end of the 16th century an account of the 'whole quantitie of the Earth' was published in a navigational treatise entitled *Certaine Errors in Navigation Detected and Corrected,* where author Edward Wright stated,

> *'And albeit the Globe of the earth and water, compared with the sphaeres of the starres, as it were a centre or prick; yet being considered by itself, it conteineth in the greatest (meridional) circle 6,300 Spanish leagues.'*

Wright went on to define the degree along the meridian, which averages 69.12 miles, as 17 and a half Spanish leagues, each league then being comprised of 20,000 'feet' of 1.042725 feet, or 10,000 sacred cubits of 2.08545 feet (this latter unit forms the ten millionth part of the polar radius). From Wright's information the meridian circumference was clearly intended to be 24,883.2 miles. Even more relevant, the fact that 6,300 Spanish leagues are equivalent to this measure makes the intended polar radius, via the by now familiar $^{63}/_{10}$ conversion, exactly 1,000 leagues, or 3,949 $^{5}/_{7}$ miles, the ancient figure.

Wright's account confirms that the both the values and the techniques of ancient geodesy had survived into the 16th century. It is directly clear that the intended conversion from polar radius to meridian circumference invoked the

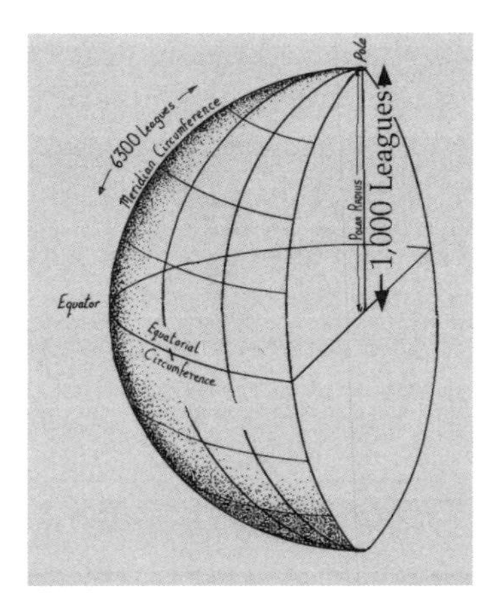

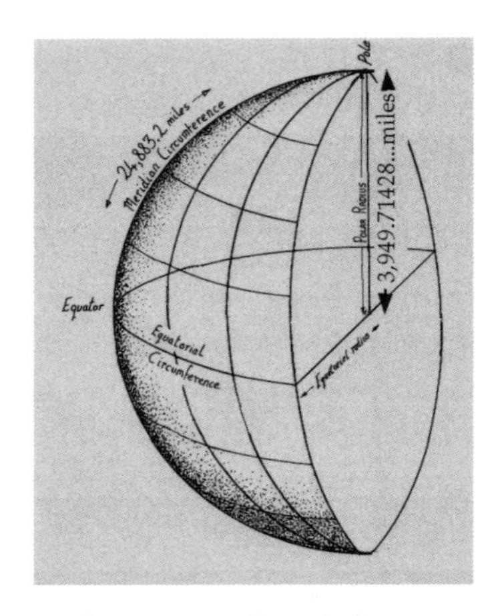

Figure 2.4. The measures and techniques of ancient geodesy survived until the seventeenth century. The dimensions on the left are from a late 16th century navigational treatise, those on the right the values of ancient geodesy. They are exactly equivalent, a 'Spanish league' being 10,000 sacred 'Egyptian' cubits. Exactly 1,000 Spanish leagues define the polar radius.

value of $^{63}/_{10}$, *pi* being enlarged to 3.15 to account for the non-circular shape of the meridian circumference.

Wright continues in his account, and concerning the degree of the meridian, the 360th part of the circumference, he writes,

'it is found by navigation at sea, and also by travel on land, that the two foresaid points are distant from each other 17 leagues and 1/2: of which leagues, each one containeth 4,000 pases, each pase 5 foote, every foot 16 fingers and every finger 4 grains of barley.'

From this account, we may identify the pace as 5.213622857 Roman feet, whence a half-pace or 'step' becomes 2.606811429 Roman feet, a value we shall return to in the following two chapters, when considering the origins of the megalithic yard.

One must conclude from Wright's account that accurate values for both the size and shape of the earth were known before the Renaissance. During the Dark Ages in Europe, assisted by the Moorish presence in the Iberian peninsula, the Arabs alone nurtured the flame of enlightenment in the sciences. In the 8th century Al Mamun, under the auspices of Caliph Al Rashid and the Islamic wisdom school, had attempted to ascertain the length of a degree of arc on the Earth's surface, evidently in ignorance of the existing and more accurate figures available from antiquity. Curiously, he appears not to have had access to the ancient geodetic figures which, seven centuries later, turn up in

Wright's treatise, based on the 'Spanish' league, surely suggesting an Iberian (i.e. Islamic) origin for the measure.

For seven centuries after Al Mamun the world slumbered in matters geodetic, and only awoke when the alarm clock of opportunity rang to announce the Renaissance. A vigorous expansion of world trade and empire building required better navigation and more accurate artillery. During the Elizabethan era, huge advances were made in navigational science, spurred on by the threat of Spanish and French invasions.

The Modern Quest

During recent centuries, the modern figures for the various radii and circumferences of the ellipsoid Earth have been honed to a fine level of precision by modern geographers and scientists. It was Isaac Newton who catalysed this expansion of knowledge, because of his need to know the size of the earth in order to ascertain its mass, a quest which was to precipitate our modern age.

Having exhausted himself researching ancient accounts of the size of the earth, and the units of length used by the ancient Egyptians and Jews, Newton came to focus on the length of one such unit, the sacred cubit. His studies had lead him to believe that this unit was *a simple fraction of the earth's polar radius,* the ten-millionth part.

The man who was to provide Newton with the information he needed to estimate the length of the sacred cubit, and thereby the size of the earth, was an Oxford professor of astronomy, John Greaves. In 1637, he had engraved some remarkably accurate measuring rods, and journeyed with these to the Pyramids. His measurement of the King's Chamber, at 34.38 by 17.19 feet showed its groundplan to be a rectangle 20 by 10 cubits of 1.719 feet, the so-called royal cubit, known to Newton as the *Memphis* or '*profane*' cubit.

The sacred cubit was known to be proportioned to the royal cubit by the ratio 6:5 *(see Fig 2.2)*, making the sacred cubit 2.063 feet, a value Newton used to obtain a figure for the size of the Earth. In his *Dissertation upon the Sacred Cubit,* Newton estimated its length as lying between 2.0605 and 2.07824 English feet, and used a value of 2.063 feet to obtain the radius of the earth. This brought him to within 42.5 miles of the actual length of the polar radius, a figure not quite good enough to balance his equations. Had Newton multiplied ten million Egyptian sacred cubits (2.0736 ft) by the metrological *pi* fraction $^{176}/_{175}$, he would have obtained the exact ancient value for the polar radius of the earth, 3949 $^{5}/_{7}$ miles.

Unsatisfied with his result, Newton decided to calculate the size of the earth for himself. Using the first reflecting telescope, his own invention, he had noticed that the planet Jupiter bulged at the equator, and theorised that this would also be true for the earth, making the equatorial radius larger than the polar radius. Consequently, he figured that the length of a north-south line covering a single degree of arc of latitude, the so-called meridian degree, on the earth's surface, would be shorter at the equator than as one approached either of the poles. To complete his estimate of the size and hence mass of the earth, Newton urgently needed to know by how much the length of a degree of latitude changed between the equator and pole.

Newton's hypothesis concerning the earth's bulge led him to believe that it was now irrelevant to glean the earth's size from ancient accounts of the polar radius, multiplying this by twice *pi* in order to find the meridional circumference. He clearly believed that the ancients had not known about the earth's bulging equator. It is reasonable to conclude that Newton had not read Wright's treatise, which showed irrefutable evidence not only that the non-circular shape of the earth was known by the ancients, but had also been effectively dealt with. It was enshrined both within the measures themselves and in the approximation to *pi* ($^{63}/_{20}$) utilised to ascertain the meridian circumference from the polar radius.

Because of Newton, the link with the ancient traditions was severed. Thenceforth, geographers began again from scratch in order to ascertain the size of the earth, finally obtaining the dimensions which now afford us the luxury of comparison with those of the ancients. Pioneers of precision, ignorant or dismissive of ancient geodetic knowledge, they considered themselves responsible for stocking an empty larder, moving onward from an ignorant past, utilising improvements in technology to solve the problems of measuring angles and lengths accurately. Using *science*, in other words, where previously they assumed there had been just confusion. This was a wholly erroneous perspective, as is now clear. In truth, modern geodeters have merely re-invented the wheel, albeit a non-circular and, more recently, a metric one.

The Enterprising Jean-Luc Picard

Newton's saviour in his quest was to be Jean-Luc Picard, a French astronomer, who in 1671 accurately measured a base-line across the Earth's surface of 80 miles length, an arc of 1.2 degrees, between Amiens and Malvoisine, in Northern France. The technology of theodolite telecopes was only just becoming available - indeed, Picard's survey was the first known performed with a telescope. It will not have escaped the more observant reader that Jean-Luc Picard later became Captain of the *SS Enterprise*, apt casting and a most suitable reincarnation indeed!

Picard calculated that the meridian degree, at the latitude of 50 degrees north, was 69.1 miles in length, a figure not to be bettered for a century and a half, and then not significantly altered. Newton took Picard's figure and, having obtained an accurate figure for the mean radius of the Earth, this led him on to discover the mass of the earth and thence to the laws of gravity and motion. The rest, as they say, is history. The Mechanical Age was born.

Serendipity played its part in Picard's choice of location for his base line. Fifty degrees north latitude just happens to coincide with the location of the average value of the meridian degree of arc. Multiplying Picard's figure by 360 thus gives a meridian circumference of 24,876 miles, a figure only seven miles short of the canonical figure of 24,883.2 miles, an average meridian degree of 69.12 miles.

The Airy Spheroid

Improved measuring techniques slowly led to more accurate figures for the size of the Earth. Sir George Airy, the Astronomer Royal in 1830, computed the flattening of the poles to be one part in 299.3, since when generations of trainee geographers and surveyors have sniggered at the merest mention of the 'Airy spheroid' in lectures. It has become known as the National Projection, and was adopted by the British Ordnance Survey and remains the best fit for the shape of the Earth around the British Isles. The National Projection gives a meridional circumference of 24,883.1284 miles, within 378 feet (99.998%) of the ancient value!

Colonel A.R.Clarke's spheroid was calculated in 1866 and lead to the most accurate surveying of the planet ever undertaken in modern times using theodolites and tapes. The Clarke figure of flattening, one part in 295, was used for all the maps of North America, India and Africa, whilst the British Admiralty charts remain constructed on the basis of a figure for the Earth known as the Clarke 1880 figure. Here, Clarke ascribed 3,949.573 miles to the polar radius and suggested that the true equatorial radius lay between 3,962.117 and 3,963.42 miles. The circumference of the equator became 24,898.81 miles (taking the average equatorial radius), whilst the meridional circumference became 24,882.31 miles. The flattening of the poles became one part in 293.5.

In 1924, the *International Ellipsoid of Reference (IER)* was formulated, and in 1984, the *World Geodetic Survey* formulated the latest values as WGS 84. Figure 2.4 (*overleaf*) illustrates the chronological order of this modern quest to ascertain the size and form of the earth, now referred to as the *geoid*. These accurate modern dimensions for the size and shape of our planet enable us to make a comparison with those used by the ancient world.

STANDARD	POLAR RADIUS	EQUATORIAL RADIUS	FLATTENING	MEAN
DELAMBRE 1806	3955.9 MILES	3967.7 MILES	1/334	
AIRY 1830	3952.4	3965.6	1/299.3	
CLARKE 1866	3949.573	3962.96	1/295	
CLARKE 1880	3949.573	3963.029	1/293.5	
IER 1924	3950.109	3963.409	1/297	
GRS 80 1980	3950.010	3963.231	1/298.7580	3958.7
WGS 84 1984	3950.076	3963.298	1/298.7581	
ANCIENT GEODESY	3949.7142...	3963.42857	1/286.53	3958.6909

Figure 2.5 Re-inventing the Wheel. The modern quest to determine the size and form of the earth has produced a succession of technological advances which, most recently, have culminated in the latest satellite mapping techniques. The resulting definitive figures all lie within half a mile of the traditional values of ancient geodesy, a canonical system connecting the principal dimensions of the earth through simple fractions. It is only by comparing the modern figures with those of the ancient world that the accuracy of the latter 's knowledge of the size and shape of the planet becomes astonishingly clear.

From such a comparison, the accuracy and elegance of the ancient system emerges. Unlike the modern values accorded to the principal dimensions of the earth, which enjoy no easy relationships to either each other nor to the units of length in common use and by which they are expressed, those used in the ancient geodetic system *are all connected through exact fractions*, facilitating derivation of one value from another. Indeed, the whole system is coherent and readily memorable once the underlying system is understood.

The Equatorial Circumference

(Modern figure, 24,902.14 miles; ancient figure 24,902.94857 miles)

Because the earth is an ellipsoid, of all the great circles that can be drawn across the globe, the only true circle is the equatorial circumference, whose ancient geodetic length was 24,902.94857 miles, (131,487,568.5 English feet). When divided by twice $^{22,698}/_{85^2}$, a very accurate approximation to *pi*, this gives an equatorial radius of 3963.42857 miles, a figure almost identical to that published in *The Encyclopedia America*.

As Michell puts it *(Ancient Metrology, page 29)*, 'whoever first established these canonical figures for the earth's dimensions knew the value of *pi* to a greater accuracy than five places of decimals. And the geodetic measurements in which this is demonstrated are not only accurate in themselves, but are brought together under one system of proportion and number in a way which shows very different preoccupations from those of modern science.'

The ancient value for the equatorial circumference relates to the meridian circumference through an exact fraction, $^{1261}/_{1260}$. Furthermore, this ancient geodetic value for the girth of the earth factorises to give a most interesting 360,000 x 365.2432 feet, a combination which suggests both astronomical and geometrical intent in the choosing of this unit, for there are 365.25 days in the calendar year and 365.2422 days in the tropical 'solar' year.

The English foot thus curiously fits into the slipper of the requirements of a scientific earth dweller. An astronomer wishes to divide the sky into the number of days relevant to the calendar/solar year, whilst a surveyor or astrologer requires that the circle is divided into 360 degrees, which makes calculations in a world without electronic computation much easier, as 360 factorises into a vast range of fractional sub-divisions, its prime factors being 2 x 2 x 2 x 3 x 3 x 5.

The various forms of the Egyptian 'solar' calendar confirm that this was done, as 360, 365.25 and 365.242 days were all used to satisfy various state, civic and astronomical calendar requirements. Theon of Smyrna even illustrates the division of the sky into 365.25 parts as a circle so divided within a treatise on the calendar. The method was apparently applied in ancient China, where the equator is shown as the circle to be similarly divided. These ancient records of the division of the equatorial circle suggest that a decision may have been taken, before the history of our epoch began, to adopt the figure of

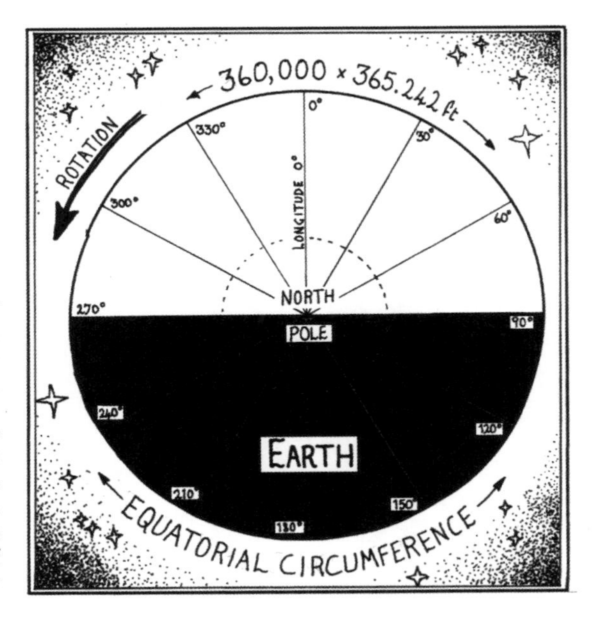

Figure 2.6 High Noon at Greenwich - a polar view of the earth. The canonical value of the equatorial circumference in ancient geodesy is 24,902.94857... miles, (131,487,568.5 feet), which is 365.2432 x 360,000 feet, implying an astronomical basis for the foot. If the true value of the solar year is substituted, then the circumference equals 365.242199 x 360,000 feet, or 24,902.8772 miles (131,487,191.6 feet), this figure lying within just 380 feet of the canonical value, and within 2000 feet of the modern figure listed in the Smithsonian tables at 24,902.6 miles (131,485,680 feet).
Approximating the length of the year to our present calendar figure, 365.25 days, the equatorial circumference becomes 24,903.40909 miles, radius 3963.500 miles. This value, and the slightly smaller value of 24,902.94857..miles was identified by John Michell in *Ancient Metrology*.

131,490,000 feet (24,903.409 miles) as the ancient geodetic value for the equatorial circumference of the Earth. If so, this decision would suggest the origin of the English foot, for 360,000 x the solar year of 365.242199 days would equal the equatorial circumference, *in feet*, to within 376 feet, or 99.9997%.

The ancient geodetic origin and hence importance of the English foot, sometimes referred to as the *geographic foot*, is revealed as being the root measure for the entire edifice of ancient metrology.

The Mean Radius of the Earth

(3,958.6909 miles)

Michell also showed that the mean radius of the earth was defined by the ancient geodeters as being the meridian circumference divided by $^{44}/_{7}$, another familiar close approximation to 2 x *pi*. This gives 3,958.6909 miles for the mean circumference, 'a figure as good as identical with the figure given in the latest *Encyclopaedia Britannica*, 3,958.7 miles'. To convert the polar to the mean radius, the fraction $^{441}/_{440}$ is employed. This fraction is another essential clue in the unravelling of ancient metrology, as will become evident shortly.

The mean Earth radius gives that value needed to calculate the area, volume, density and hence mass of the geoid, assuming it to be a perfect sphere. It was the mass of the earth that Newton needed in order to discover the gravitational constant.

These dimensions, and their ingenious interconnecting fractions, illustrated below, form the basis of the ancient geodetic system.

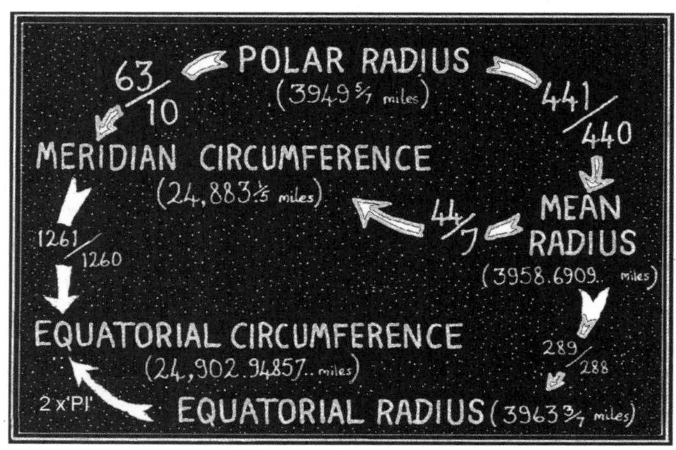

Figure 2.7 The principal dimensions of the earth, and their interconnecting fractions, form the numerical structure of ancient geodesy. These conversions are well suited to mental calculation.

The Siting of Ancient Temples

A further surprise in any study of ancient surveying techniques comes indirectly, from a study of the siting of many of the most important monuments erected by various civilisations in antiquity. The Pyramid complex at Giza is located at a latitude of almost 30 degrees, a third of the quadrant from equator to pole. If this be thought of as a chance affair, then a look at the sitings of the mighty temple complex at Thebes (Heliopolis), the geodetic capital of Old Kingdom Egypt and Avebury henge (in England) suggests otherwise, for these are placed exactly at $^{2}/_{7}$ths and $^{4}/_{7}$ths of the quadrant distance from equator to pole, at 25.7 ° and 51.42°.

Thebes was the capital of unified Egypt and the Temple of Amun was the geodetic centre from which all distances were calculated. In the case of Avebury, the required latitude band $^{360}/_{7}{}^{\circ}$ *runs right through the middle of the henge and its ruined stone circle,* the largest in the world. The Grecian city also named Thebes is however not precisely placed at the $^{3}/_{7}$ point, yet this location holds vital information concerning the system of ancient geodesy. At this latitude, the meridian degree is 364,126 feet, or 360,000 x 1.0114612 feet. This latter measure is $^{176}/_{175}$ squared, a fraction which we have already seen to be important in the system of ancient metrology.

These sitings suggest that ancient surveyors, across many cultural barriers and time-frames, were able to calculate the latitude of a location to extreme accuracy. Whilst their techniques remain obscure, the sites remain as does their geodetic system. Assuming that the choice of latitude for these sitings was deliberately taken, it follows that the ancient surveyors could have estimated the meridian circumference of the earth just as, in 1671, Picard did.

Measuring Latitude

To measure latitude one has to first understand that the Earth is spherical, then measure the culmination angles of stars very accurately indeed. We know of no mechanism or artefact from the ancient world that would allow such an accuracy to be obtained. Measurement over many miles can of course be undertaken using rods or ropes marked out in any arbitrary unit of length, without reference to the sky. But the instant the bearing (azimuth) of a surveyed line is needed, then reference must be made to the sky. Even a neolithic burial with the body laid east-west requires astronomical knowledge. And having discovered that the latitudes and units of length incorporated into major temple sites of the ancient world are intimately related to important divisions of the meridian circumference, the conclusion must be drawn that a precision science of surveying, astronomy and metrology existed in antiquity.

Caesar's Druids cited at the very beginning of this chapter must have been measuring latitude in order to fulfil their teaching duties, unless they had inherited their knowledge concerning the size of the earth from elsewhere, or from an earlier culture. The Druids came two thousand years after the culture which built the stone circles in Britain and although we have no written records from that period, Professor Thom showed that the accuracy by which their fixed observations were made could "split a minute of arc, that's better than a modern surveying theodolite."* What then could the stone circle builders and the later Druids not do that Picard and the 18th and 19th century surveyors could? This is an important question, because the present historical model fails to square with the geodetic facts emerging from the ancient world.

Then there's the additional problem of determining longitude, considered impossible to measure accurately without a chronometer. No one has ever been fortunate enough to find a Rolex in an ancient burial chamber.

Measuring Longitude

To measure the difference in longitude between two sites requires that the same moment in time is known and observations of the sky taken at both locations simultaneously. The development of accurate chronometers facilitated such observations, but there were techniques available in the ancient world, which are known to have been applied, involving the measurement of angles. The simplest example is the observation of the first moment of the earth's shadow across the full moon during a lunar eclipse. This will be observed at a different location in the sky depending on the longitude of the observation. In Egypt, an eclipse beginning when the moon is due south, will, in Spain, be seen some thirty degrees further over to the left, in the south-east. It follows that Spain is thirty degrees west of Egypt. Using this 'clock-face in the sky' to obtain longitude once again requires the accurate measurement of angles, just as it does for the measurement of latitude and in land surveying.

Some of the earliest cuneiform tablets from Babylonia demonstrate that the required trigonometry was established prior to 2000 BC. One such tablet (Plimpton 322) lists a set of fifteen Pythagorean (whole number right-angled triangles) whose apex angles cover the range from 30° to 45°. The ancients were well nigh obsessed with the measurement of angles and research into geometry - a word which today is often taken to mean the creation and analysis of shapes, yet actually means 'measuring the world'; *Geo*-metry.

* Quoted from a BBC *Chronicle* documentary, *Cracking the Stone Age Code*. 1973

Two Revelations

The connection between the mean radius and polar radius of the earth, dimensions we know from ancient accounts to have been 3,958.6909 (3,958 $^{38}/_{55}$) miles and 3,949.7142857 (3949 $^{5}/_{7}$) miles respectively, are numerically related by the fraction $^{441}/_{440}$, as shown by Michell in *The Dimensions of Paradise*. This is illustrated below, connecting 'root' and 'standard' measures. In effect, the conversion is made between 'polar' and 'mean' measures.

In *Ancient Metrology*, Michell demonstrated that many measures were adjusted according to an exact $^{176}/_{175}$ fraction, related to the latitude of the site.* In *All Done with Mirrors*, John Neal shows how the irrational pi may be rationalised using 'diameter' and 'perimeter' measures, also related through this same fraction *(see page 20)*. Neal then went on to combine both these fractions, thereby integrating the ancient metrological system within a four column system, each column separated either by the fraction $^{441}/_{440}$ or $^{176}/_{175}$ *(see below, figure 2.8, and overleaf, figure 2.9)*.

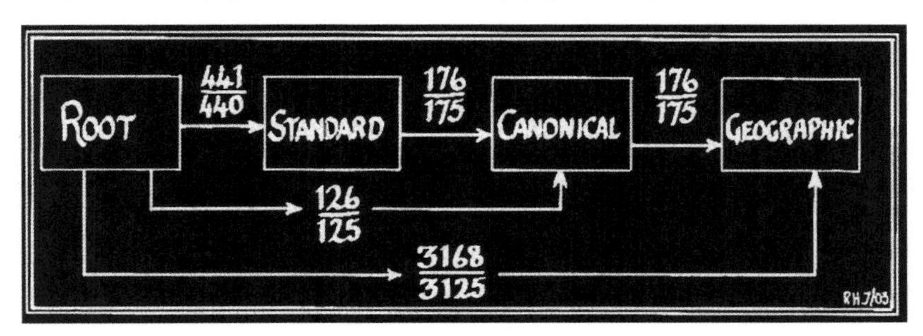

Figure 2.8 The structure of ancient metrology *(after Michell and Neal)*. The 'root' measure, always based on the foot or a simple fraction of the foot, is multiplied by the fractions above to align the measure to suit specific geodetic and mundane applications, according to the latitude.

Summary

Throughout the historical period, traces of the traditional system of geodetics and metrology have occasionally surfaced and then apparently been misunderstood or forgotten. Edward Wright in the 16th century showed that this knowledge had survived the Dark Ages, as, a century later, so did Newton, who knew of the tradition of ancient measures being related to the length of the polar axis, yet failed to decode the system and evidently had not

* The same type of measure taking a shorter 'tropical' length and a longer 'northern' value, connected by this ratio, this practice related to the number of feet in a single degree of latitude, measured at 10° and 50°, values which are 362,880 (factorial 9 or 9!) and 364,953.6 feet respectively.

THE SYSTEM OF ANCIENT METROLOGY				
TYPE	ROOT	STANDARD	CANONICAL	GEOGRAPHIC
Conversion		x441/440	x176/175	x176/175
GREEK				
foot	1.000000	1.0022727	1.008	1.01376
cubit (x 3/2)	1.50	1.5034091	1.512	1.52064
mile	5000	5011.363636.	5040	5068.8
EGYPTIAN				
foot	1.142857..	1.14545..	1.152	1.158582857
cubit (x 3/2)	1.7142857..	1.7181818..	1.728	1.737874285
mile			5760	
ROMAN				
foot	0.96	0.9621818	0.96768	0.9732096
remen(x 5/4)	1.20	1.202723..	1.2096	1.216512
cubit (x 3/2)	1.44	1.4432727..	1.45152	1.4598144
mile	4800			
POLAR				
foot (6/7th Egyptian foot)	0.9795918	0.981818145	0.987428534	0.993071
SACRED				
foot	1.37142857..	1.3745454..	1.3824	1.3902994
cubit	2.057142857..	2.061818..	2.0736	2.0854492

All values are in ENGLISH feet. Measures employed at ancient sites or where examples exist are shown underlined. Other root feet, always connected by simple fractional ratios to the English foot, are discussed where appropriate in the text.

Figure 2.9. The measures of the ancient world, accommodated within the four columns of figure 2.8. Those shown underlined are well attested measures, examples of which exist in museums or which have been employed at important ancient sites. In effect, the matrix shown here is the combination of figures 2.2 and 2.8. Other root values for the foot, each related by a simple fraction to the root 'Greek' foot, better known as the English foot, are discussed in Appendix One, page 122. The table above is simplified from *All Done With Mirrors*, by John Neal.

known of Wright's treatise. His quest to ascertain the true values for the size and form of the earth began from scratch, yet has ultimately produced dimensions insignificantly different from the traditional ancient values and much less convenient to apply in geodetic calculations. These modern measurements do however afford us for the first time the luxury of comparison.

Caesar's quote concerning Druidic knowledge of the dimensions of the earth is now supported by evidence that a system of land-surveying has existed since prehistoric times, in many different places and within many different cultures. And while we call measures by names that link them to certain areas of the globe or to certain known civilisations, the emerging picture is that, as subdivisions of the principal dimensions of the earth, these measures form part of a common metrological legacy dating from prehistory.

What was it used for? We have shown that the fractional nature of the system allows for an easy conversion between units, latitudes and geometric shapes in a time before the electronic calculator and the computer. In effect, once the root measure of this system is known, together with the principal dimensions of the earth, the required conversions suited to a particular application or location may readily be derived. The resulting measures and dimensions related a monument, edifice or way-stone distance directly to the location on the planet upon which it was placed. This is applied sacred geometry, a high consciousness meld of science, art and magic.

The English foot lies at the root of this system. Most mysteriously, when John Greaves had completed his measurements in the King's Chamber within the Great Pyramid, to reveal an accurate definition of the length of the royal cubit, he engraved a copy of the English foot, from his Guildhall standard, on the wall. Underneath he wrote, 'To be observed by all nations'. Unless Greaves had been seized with an uncharacteristic outburst of nationalism, this single act suggests that he may have been privy to some arcane knowledge, the subject of this chapter, which was not publicly disclosed following his trip to Egypt.

We can be much more certain that Petrie came very close to discovering the system of ancient geodesy. In the 1910 edition of the *Encyclopaedia Britannica,* and in his 1934 book, *Measures and Weights,* he stated that certain measures varied by the 170th and the 450th part. Neal remarks, in *All Done with Mirrors* 'Doubtless, he would have appreciated these small refinements (of the 175th and 440th parts) to his observations'.

Michell and Neal have now recovered the major part of the system of ancient geodesy and metrology, but this does not answer the questions posed at the very beginning of this chapter. For that, we must now turn to the moon.

Figure 3.1 The Moon. The phases of the moon take an average of just over $29^{1}/_{2}$ days to complete, a period termed the lunation cycle. The lunation begins with the new moon. Here we see the moon waxing, just over a third through the cycle, some 11 days after the new moon and four days before the full moon. Many cycles on the earth are synchronised to the lunation cycle, including the tides and the human menstrual cycle. There are just over $12^{1}/_{3}$rd lunations (12.368...,almost exactly $12^{7}/_{19}$ths) in the solar or calendar year of $365^{1}/_{4}$ days, the photograph showing the moon at $^{7}/_{19}$ths through a lunation.

Chapter Three

Measurement and the Moon

'The heavenly bodies are beyond reach, everywhere present yet impeturbable; they form the basis of cosmo-logical frameworks. The cycles of the sun, moon, planets and stars are also by far the most dependable regularly recurring natural events'

Professor Clive Ruggles, Chair of Archaeoastronomy, Leicester University*

The great metrologists of the past have assured us that time measurements preceded all the other measures. In ancient times, the motions of the luminaries were the cosmic clock, their rhythms and the daily rotation of the sky providing the tick and the tock for the accurate measurement of time, through the measurement of angle. Professor Hogben, author of *Mathematics for the Million,* the best selling popular mathematical book of all time, describes many ancient applications which relied on treating the vault of heaven as a clock-face rotating around a central and fixed earth. He shows how, by making observations of lunar eclipses, the problem of longitude was solved three millenia before Harrison's chronometer. The moon, as the fastest moving body in the heavens, provides, as it were, the 'second hand' in such a cosmic time-piece. We must not therefore be surprised that *mensuration,* our modern word for the accurate measurement of time, length, area, or volume, relates directly to the moon, for its root, *mensis,* means month.

In the Indo-European group of languages, from India to the western Celtic fringes of Europe, the word *men, man* or *maen* means 'stone'. Thus, in Cornish, *men-an-tol* means the stone with the hole, and in Welsh, *rhos maen* means stone of the moor; whilst in southern India, *mansmai* means oath-

*Quote from *Megalithic Astronomy* (BAR 1983), introduction. ISBN 0 86054 253

stone, *manloo* stone of salt and *manflong* a grass-covered stone. Returning to western Europe, *menhir* means long-stone, whilst *dolmen* means table-stone. So, laid out on an etymological table in front of us we find moon, month, measurement and stone all connected.

While the etymological connections between lunar cycles and measurement remain, the cultural priorities have subsequently changed. Modern man no longer cares much about the moon, nor why it might have come about that the regular time cycles of the moon once led to the measurement of distances on the earth. But is there a link between the moon and megaliths, and a connection with the unit of length by which they were laid out? Is there a *numerical* connection between the moon and measurement? We begin answering these questions by showing just how ancient is the connection between the moon's cycles, its numbers, and human culture?

The principal lunar cycle affecting life on earth is the synodic lunar month or the *lunation cycle*, of 29 1/2 days duration. The regular phases of the moon are generated by two apparently independent phenomena - the combined motion of the lunar orbital period around the earth, the *sidereal month*, which takes 27 1/3 days to complete, and that of the earth's orbit around the sun. Both 'months' are discussed in more detail later in this chapter.

Prehistoric Moon-lore

The link between the moon and human culture is explored by anthropologist Dr Chris Knight, who in his book *Blood Relations - Menstruation and the Origins of Human Culture* (Yale 1991), investigates the cultural connections between the moon, menstruation and the lunation/tidal cycle. He notes that,

'..*the human female menstrual cycle - virtually alone amongst primate cycles - is a body-clock with precisely the correct average phase-length to enable lunar/tidal synchrony to be maintained.*'

Knight's theory, that the human reproductive cycle became synchronised with the lunation cycle through cultural forces, holds wide social implications. Of all primates' reproductive cycles, the human is the only one to have synchronized so precisely with the lunar phases. The cultural link between moon, blood and man is suggested by the repeated use of red ochre in prehistoric cave art and the use of white quartz and red ochre around burials.

The famed *Venus of Laussel* holds with one hand her gravid belly and, in her other, a crescent shaped horn scored with thirteen or fourteen notches. Even in those far distant times, the human fertility cycle had apparently been connected with the menses, the month and thence the moon, specifically, the lunation cycle.

Mircea Eliade suggests that human monitoring of the lunation cycle is astonishingly ancient,

'The fact remains that the lunar cycle was analysed, memorised, and used for practical purposes some 15,000 years before the discovery of agriculture.'

Alexander Marshack, who studied tally and score marks on prehistoric artefacts, went even further, claiming an unbroken sequence of lunar observation going back over 30,000 years.

'The evidence is neither sparse nor isolated; it consists of thousands of notational sequences found on the engraved 'artistic' bones and stones of the Ice Age and the period following, as well as on the engraved and painted rock shelters and caves of Upper Paleolithic and Mesolithic Europe.'

Add to this the lunation numbers found carved on stones at Knowth, in Ireland, and in the number of stones chosen for the Sarsen circle at Stonehenge, 29 full width and one half width stone (Stone 11 on the HMSO plan), and one can safely suggest that lunar observation has formed a long unbroken strand in our cultural development. Only recently, since the introduction of artificial lighting, has this link been severed. Night no longer means dark to most city dwellers, and few people now know or care about the current phase of the moon, the state of the tide nor when the moon will rise and set.

Stone Circles and the Megalithic Yard

Littered across Western Europe, large numbers of stone circles and standing stones were erected in the Neolithic and Bronze Age period. The unit of length with which many of them were laid out was discovered by professor Alexander Thom, who surveyed several hundred stone circles between 1937 and 1976. Following extensive statistical analysis by two of the most renown statisticians of the day, Broadbent and Hammersley, Thom's results were later confirmed by 'that most ferocious of mathematicians' David George Kendal. Thom suggested that there was 'a presumption amounting to a certainty that a definitive unit was used in setting out these rings', and proposed to call this unit the Megalithic yard. Its length was 2.72 feet (0.829 metres), and it was employed in the layout and construction of many stone circles and other megalithic monuments.

Curiously, Alexander Thom never connected 'his' unit with lunar cycles, and the numerical connection between the moon, stone circles and the unit of measure used in their construction has waited over thirty years to be revealed.

These enigmatic structures do indeed link the moon with measurement. The much disputed megalithic yard can be shown to connect stone circles with the lunation cycle and, crucially, the system of ancient metrology and geodetics.

The Lunation Cycle

The lunation cycle is the period of the moon's phases. To return to the same phase takes between 29 and 30 days. On any given day, the Moon displays the same phase to all inhabitants of our planet and during every 30 day period, there will always be one full moon and one new moon.

It is impossible to over-estimate the practical and social importance this cyclic event commanded on everyday life prior to the widespread introduction of artificial lighting. The amount of light available during the night reaches its maximum at the full moon, facilitating travel, hunting and social interaction, whilst the new moon, some 15 days later, presents us with an almost totally dark night-time, inhibiting movement. Between these two profound astronomic events, the two 'quarter' moons, which are visually half the moon's disc and visible even in full sunlight, provide an instant reminder as to which phase of the lunation cycle the Earth is presently experiencing. The rhythm is thus new moon, first (waxing) quarter, full moon, third (waning) quarter, followed inexorably by a return to the new moon. The long-term average duration of this cycle is 29.53059 days.

Understanding that there are 29.53059 days in a lunar month appears to be a giant leap from scoring 29 marks on a reindeer bone, although gleaned from the pages of an astronomy text-book it probably means far less to a modern reader than those integer score marks on a bone did to its original owner. But how might a prehistoric culture have been able to assess this *average* value for the period of the lunation cycle? The techniques have been discussed elsewhere, and suggest that it is entirely reasonable to assume that the task was solved. Proof that it was solved comes from another route and other evidence.

To begin to understand quantitatively the link between the moon and measurement one needs to understand time cycles, the calendar, and therefore numbers. The assumption that ancient man was able to count to high numerical values has been widely questionned by archaeologists and prehistorians alike, because they have been unable to find any physical proof or artefact that confirms this. We are now in the position of being able to offer proof, with the moon, more specifically its lunation cycle, and stone circles as the artefacts in question. The time periods involved in the lunation are tabulated opposite.

The Astronomy of the Lunation

Solar (seasonal) year = 365.242199 days

Lunation period = 29.53059 days

Difference between the solar year and lunar year = 10.87512 days.

(the lunar year is 12 lunations, 354.367 days)

Difference expressed as fraction of a lunation = 0.368267 (The 'Silver Fraction').

Number of lunations in one solar year = 12.368267.

The Lunation Cycle and Stonehenge

To discover the cosmological framework underpinning soli-lunar recurrence, it is helpful to convert this last number into fractional notation. It is almost exactly $12\,^{7}/_{19}$ths, which is $^{235}/_{19}$. It follows that in 19 solar years there will have elapsed 235 lunations, this cycle being known as the Metonic cycle. Attributed to a 4th century BC Greek mathematician, Meton, knowledge of this cycle may be much older. The 19 year soli-lunar recurrence, which occurs within two hours of exactitude, is the most accurate of soli-lunar repeat cycles, knowledge of which facilitates the design of accurate calendars.

In 40 BC, Diodorus wrote,

'Hecataeus and some others tell us that opposite the land of the Celts there exists in the ocean an island not smaller than Sicily, and which, situated under the constellation of the Bear, is inhabited by the Hyperboreans.. (who).. honour Apollo more than any other deity. A sacred enclosure is dedicated to him in the island, as well as a magnificent circular temple adorned with many rich offerings. Apollo visits the island every 19 years.'

Nineteen slender dressed bluestones, in a horseshoe shape, adorn the central area of Stonehenge, a monument which is certainly 'a magnificent circular temple', and demonstrably functions as a soli-lunar and hence calendrical device. There are other important stone circles which contain 19 stones, the Merry Maidens and Boscawen-un in Cornwall *(fig 1.3)* perhaps being the best known. Was knowledge of the Metonic cycle known by the architects of Stonehenge? To attempt an answer to this important question, we first investigate the relationship between the dimensions of the Aubrey circle and the outer sarsen circle diameter. Whichever of the many published values* one takes for

*Taken from Petrie's 1877 survey of the sarsen circle and Thom's 1973 survey, these diameters are given as 104.27 and 283.6 feet respectively, and form a ratio within 0.2% of $^{7}/_{19}$.

the diameters of these two principal circles at Stonehenge, the ratio between them always falls within 0.5% of $^{7}/_{19}$. What are we to conclude from this, that such a fact is mere coincidence?

The next relationship to be explored is that revealed by inquiring how many $^{7}/_{19}$ths of a lunation period fit into the lunation period itself. The answer is the reciprocal of $^{7}/_{19}$ths, or $^{19}/_{7}$ths. As a decimal number, this is 2.7142857.., more precisely,

The number of 0.368267 lunation periods in one lunation = 2.71542.

Here we make contact with a familiar number in the study of megalithic science, for rounded up to two decimal places it is numerically identical with Alexander Thom's megalithic yard, of 2.72 feet. Might then the megalithic yard be connected with the lunation cycle?

The Lunation Cycle and the Megalithic Yard

It is indeed fortunate that Thom undertook all his measurements in English feet. This offers a direct numerical insight into the metrological meaning of the unit. If Thom's megalithic yard, at 2.72 feet, is taken to represent the lunation period, then the 'silver fraction', 0.368 of a lunation, represents the differential period between the lunar and solar year. It is 1.00096 English feet in length, which is just twelve thousandths of an inch over the standard measure. This is so close to the root measure that the following proposition can be made.

Proposition One: If the English foot is made to represent the difference in time between the lunar and solar year (10.875119 days), then:-

The megalithic yard, based on the astronomy of the lunation cycle of 29.53059 days, becomes 2.71542857 feet in length.

And: *foot : megalithic yard = differential : lunation period.*

The difference between Thom's statistically derived unit and the lunation derived unit, which I have named the *Astronomical Megalithic Yard* (AMY) is just a twentieth of an inch in nearly 33 inches. [1.4 mm in 829 mm]. But is this too just a coincidence? What we now need is some solid evidence for the use of this lunation derived unit elsewhere within ancient or prehistoric culture. In feet, the AMY is an exact 19.008 divided by 7, and we have met septenary fractions in previous chapters, relating to the polar radius. If the AMY formed part of the system of ancient metrology then we might next reasonably ask whether this unit is also related to the dimensions of the earth.

The Astronomical Megalithic Yard in Ancient Geodesy

In chapter two, we saw how, in ancient geodesy, the polar axis was taken as being $3,949\frac{5}{7}$ miles. When we convert the ancient value into Astronomic megalithic yards, we are astonished to discover that the result is 7,680,000, a canonical number $2^{12} \times 3 \times 5^4$. That such an exact value should define the Earth's principal dimension, from a unit derived from the lunation cycle, a time cycle, is both deeply satisfying and profoundly mysterious to a cosmologist, and exciting for any researcher into megalithic culture.

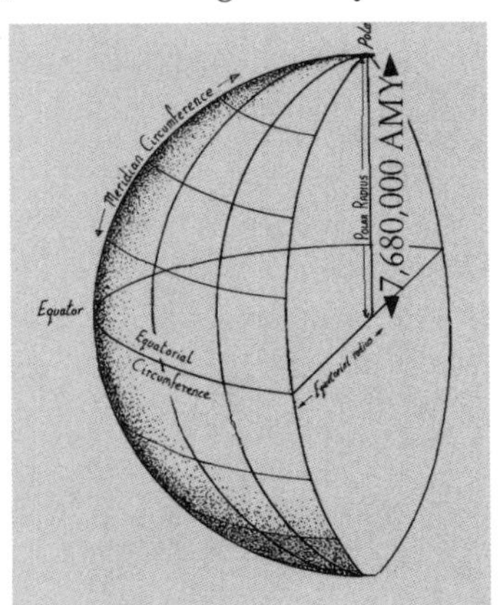

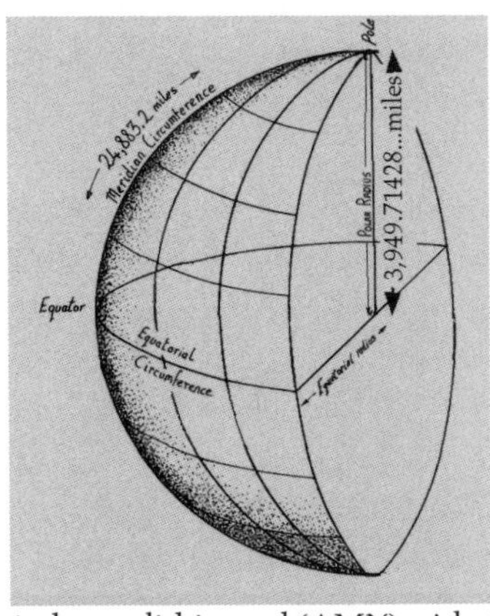

Figure 3.2 Metrological equivalence of the astronomical megalithic yard (AMY) with the values of ancient geodesy. The polar radius is $3,949\frac{5}{7}$. miles, an exact 7,680,000 AMY, linking the lunation cycle - a time cycle - to the principal earth measure. Compare this with Fig 2.4. The moon's radius is an exact 2,100,000 AMY, or 1,080 miles. [*One AMY = 2.71542857 feet*]

An equally satisfying 2,100,000 AMYs is to be found on the moon, whose polar radius in miles is 1,080 miles. In sacred geometry the ratio of the size of the moon to that of the earth is often taken as 3 : 11, but here we see an astonishingly accurate 210 : 768, defined by using figures based on the AMY as the polar radius for each respective body.

polar radius = $3,949\frac{5}{7}$ miles = 7,680,000 AMY

lunar radius = 1,080 miles = 2,100,000 AMY

If we now take the *meridian circumference* of 24,883.2 miles, a figure used by the ancient geometers and one which still defines the *nautical mile* (1' of arc, or 6,082.6 feet) and the *knot* (1.152 miles/hr or 1′ of arc/hr), then a further matrix of inter-relationships becomes immediately apparent *(Figure 3.3*

*The AMY is 19.008/7 feet, and 19.008 is 6 x 3.168. Traditionally 3168 is a number identified with the circumference of sacred temples. For example, at Stonehenge, 316.8 feet is the mean circumference of the sarsen circle, and each pair of Aubrey holes is separated by 31.68 feet. 2.71542857 feet is 6/7 of 3.168 feet, demonstrating the AMY's connection with sacred geometry.

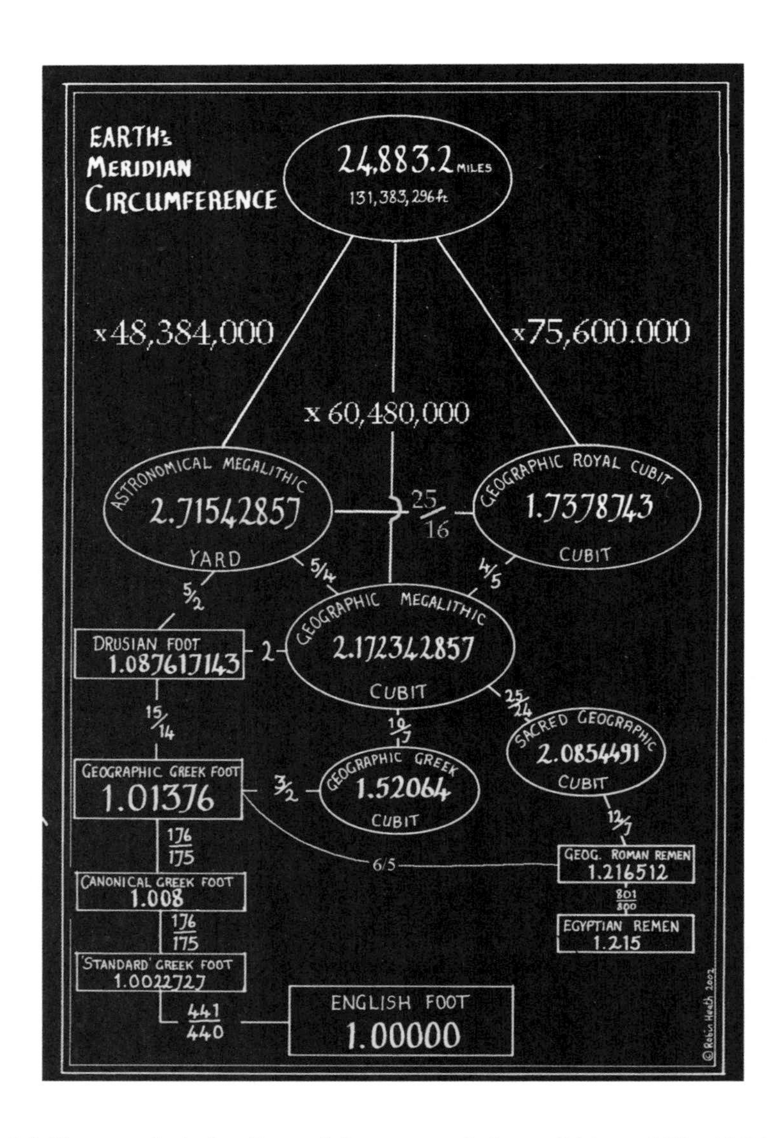

Figure 3.3 The metrological pedigree of the astronomical megalithic yard. The AMY relates to *both* principal earth dimensions, the polar radius and the meridian circumference, suggesting that the lunation, a time cycle relating the motions of sun, moon and earth, was incorporated into the geodesy of the ancient world. The diagram above shows *(lower left)* the standard route from the English foot to the Greek foot *(see also figure 2.8)*. A known measure from northern Europe, the Drusian foot, discussed in later chapters, is then a fourteenth part larger, and the AMY a further 5:2 larger still. The royal cubit and the AMY are related by the fraction $^{25}/_{16}$.

opposite) between the AMY and ancient geodesy. Further confirmation comes from Wright's 'half-pace' or 'step' measure of 2.606811429 Roman feet, *(see page 14)*, which is $^{24}/_{25}$ ths of the AMY. Note too that the geographic royal cubit, which doubles up to become a royal yard, is $4^2/5^2$ of the AMY.

Both the key dimensions of the earth integrate perfectly with the value of the megalithic yard derived from the lunation period, asserting that the AMY occupied a place within ancient metrology. The large number of decimal places quoted for the lunation period and for the ancient units of length demonstrate the precise fractions present in all these numerical operations. These calculations are neither spurious nor coincidental, they are *exact*, and their implications are monumental in understanding ancient science.

We can now make a second proposition:

The lunation cycle, represented as a length we now call the Astronomical megalithic yard (AMY), once formed part of an immensely rich interrelationship which connected, through simple numerical ratios, known historical measures from a variety of ancient cultures. Each of these measures is fractionally related to the size and shape of the earth, demonstrating that a highly sophisticated system of metrology and geodesy was being practised in prehistoric times, certainly as far back as the cultures which built the stone circles throughout Britain and northwestern Europe.

The nature of this proposition has been demonstrated here to constitute a proof by any modern scientific criteria and we are suddenly faced with an unavoidable truth. Our current orthodox model of history is shown to be plainly and absurdly wrong, so completely wrong as to throw into disarray our present beliefs concerning the capabilities of prehistoric man.

Summary

We have demonstrated the link between the moon and measurement, and in so doing, that of time cycles to linear measures. The astronomical megalithic yard takes centre stage in an 'imperturbable' cosmological framework of soli-lunar cycles, as Dr Clive Ruggles muses in his quotation at the beginning of this chapter. But it does more than this, the AMY confirms that a precision astronomy, a precision geodetic system and a precision metrology, related to the size of the planet and its satellite, was in existence prior to 3000 BC. It is, of course, 'Sandy' Thom's *Megalithic Science*.

It honours our ancestors to ponder the implications of this revealed megalithic science and then to ask how this knowledge came to disappear from the historical landscape?

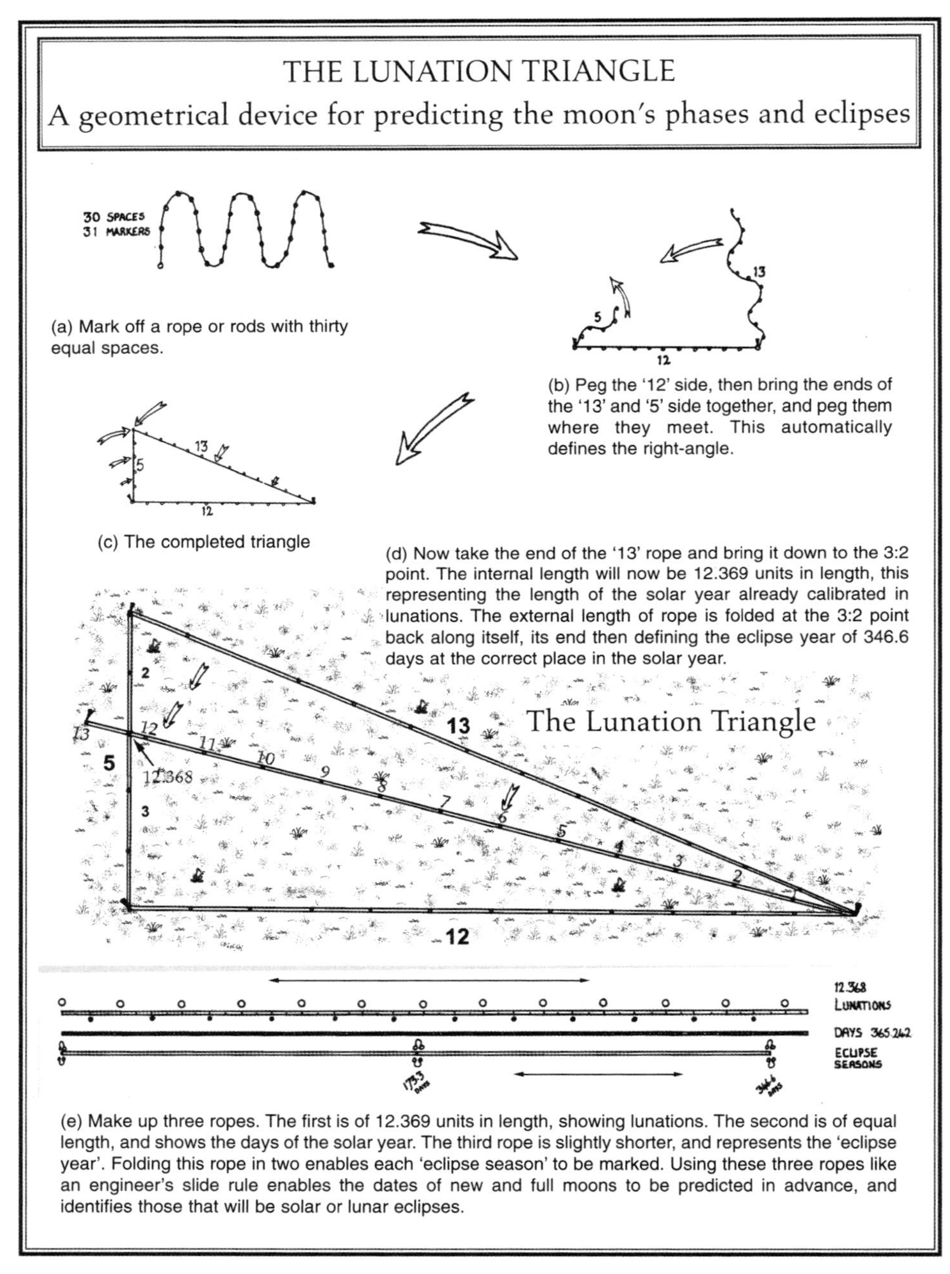

Figure 4.1 Astronomy on a rope. The lunation triangle is a simple geometrical tool that shows the dates of full and new moons, and also predicts eclipses. All that is required to generate an accurate soli-lunar calendar is a long rope marked off with 30 equal lengths and a few pegs.

Chapter Four

Stonehenge and the Lunation Triangle

'The history of research at Stonehenge may also serve, as it were, as a diminishing mirror in which is reflected in miniature the whole current of British antiquarian thought and archaeological method, from the Middle Ages to the present day.'

Professor Richard Atkinson, *Stonehenge*, 1956

My first major impetus towards original research on the megalithic culture came with the purchase of Dr John Edwin Wood's *Sun, Moon and Standing Stones* (Oxford 1978), and with perfect synchronicity a revelation arrived just a few days later. While teaching a practical electronics course to students of the Royal Aircraft Establishment, at Farnborough, I received a call from a boat-builder asking if we could provide any research into a practical tidal indicator suitable for small boat owners. The college was often at a loss for suitable project material, so I gladly put a student in touch with the boat-builder and thought nothing more on the matter. Two weeks later, the student arrived at my office door with his initial proposal, an overly complex nightmare designed to show the state of the tides. It would not have worked.

I groaned inwardly. Modelling the tides into an electronic display was evidently not going to be an easy task. The height of the tides is related to the phase of the moon, and hence the lunar month, the *lunation* period of $29^1/_2$ days, while the time of high and low water is determined, for a given location, by the position of the Moon in the sky. To build a tidal indicator, my student would have to incorporate both these rhythms within his model.

The position of the Moon in the sky is not the same thing as its phase. The 'lesser light' travels around the Zodiac, returning to the same star, in $27\frac{1}{3}$ days, a period called the *sidereal* month. During this same period the sun has appeared to move about 27° around the Zodiac. Because the moon moves about 13° a day, it therefore takes her a little over two days to 'catch up' with the sun to complete a *lunation cycle,* the time period between consecutive new moons. The lunation, or *synodic* month, is the alignment of sun, moon and earth, and on average takes just over $29\frac{1}{2}$ days to complete.

I gave my student a brief astronomy lesson, and left him to get on with a new design. Later that week, he suggested a ring of 28 LEDs and the similarity between this proposed tidal indicator and the circle of 56 Aubrey holes at Stonehenge became immediately obvious to me. The model my student was proposing to fabricate from electronic components was a copy of the Aubrey circle, itself an analogue of the astronomy of the sun and moon's movements. He knew nothing about Stonehenge yet had found the same design by studying the astronomy of the Sun and Moon.

Twenty-eight is the *minimum* number of markers placed around a circle to simulate the motions of Sun and Moon in the sky. To predict eclipses, this must be doubled, to 56, as for the Aubrey circle. My student went on to build his device which provided me with the impetus to build my own scale model of Stonehenge, the device having tracked the Sun, Moon, eclipses and lunations ever since. 28 or 56 markers placed around a circle is the only effective way to bring the motions of the Sun and Moon down onto the Earth - it forms an excellent and minimalist *design,* as astronomer Fred Hoyle discovered when he first saw what the Aubrey circle was telling him. Hundreds of these replicas of the Aubrey calendar have since been built to this design, as described in my original article in *Kindred Spirit* magazine.

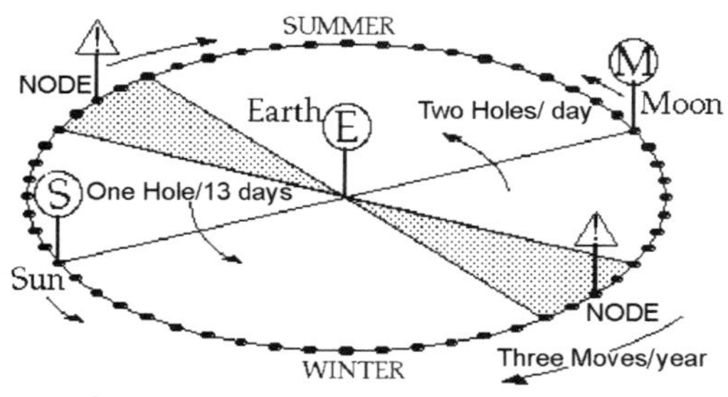

Figure 4.2 Astronomer Professor Fred Hoyle's interpretation of how the 56 Aubrey holes at Stonehenge can provide an accurate soli-lunar calendar and eclipse predictor. The device shows the date, and the position and phase of the moon. If a full or new moon occurs in the shaded area, then there will be a lunar or solar eclipse. The Aubrey circle dates from around 3,000 BC.

The 56 markers are a circular representation of the year. A Sun marker is moved two holes anticlockwise every 13 days; a Moon marker two holes anticlockwise per day. The two eclipse markers are moved clockwise three times a year. Shown here is a full moon in October, and no lunar eclipse. The 'eclipse seasons' (shaded) are shown here occuring in August and February. These rotate backwards around the calendar. and take 18.61 years to complete a circuit.

In the earliest literate cultures of India and China, the ancient astronomers and astrologers divided the circle of the sky into 28 (sometimes 27) lunar 'mansions' or *nakshatras*. This is historical evidence of knowledge of the daily motion of the Moon and her *sidereal* month period, prior to 2000 BC. At the same period the Stonehenge sarsen circle was erected, originally thirty upright stones with thirty corresponding lintels. The southernmost upright, stone 11, is half the width of the rest, this numerically indicating $29\,^1/_2$, which surely hints that the emphasis of astronomical interest at Stonehenge had shifted from the sidereal to the lunation period within a few hundred years.

The Station Stones

There are between 12 and 13 new moons in a solar year, the true figure being 12.368 lunations. The station stone rectangle, whose four corners are placed on the perimeter of the Aubrey circle, frames the sarsen circle, and its sides form an accurate 5:12 ratio. The diagonal of this rectangle is therefore 13 of the same units, completing a 5:12:13 Pythagorean triangle. The diagonal length is the same as the diameter of the Aubrey circle.

These units each turned out to be eight of Thom's megalithic yards, or 8 x 2.72 feet, thus ;

The 13 side = the diameter of the Aubrey circle (104 MY) = 282.88 feet,

the 12 side (96 MY) = 261.12 feet, and the 5 side (40 MY) = 108.8 feet.

This apparent link between the geometry and the metrology of Stonehenge greatly excited me, and led me to construct a large 5:12:13 triangle using rope and pegs. Thirty equal divisions were separated by knots, enabling me to establish the point on the shortest side where a second or intermediate hypotenuse would be near enough to the required 12.368 units in length. To my astonishment, this point occured where the '5' side became divided into the ratio 3:2, the musically harmonious major 'fifth' ratio. The construction aligned perfectly with the numerical ideas of the Pythagoreans, the cosmology espoused in the oldest Rig Veda codices, and within Hindu-Greek musical scale theory. In 1988, when this construction was first revealed to me, I named it the *lunation triangle.*

The Sarsen circle outer diameter is 0.368 that of the Aubrey circle, almost exactly $^7/_{19}$, and this 'silver fraction', as I termed it, was the *essential* component of soli-lunar and hence calendrical knowledge. Here it was to be found connecting the two main circles at Stonehenge, another impulse setting me off to delve deeper into the mysteries of this enigmatic monument.

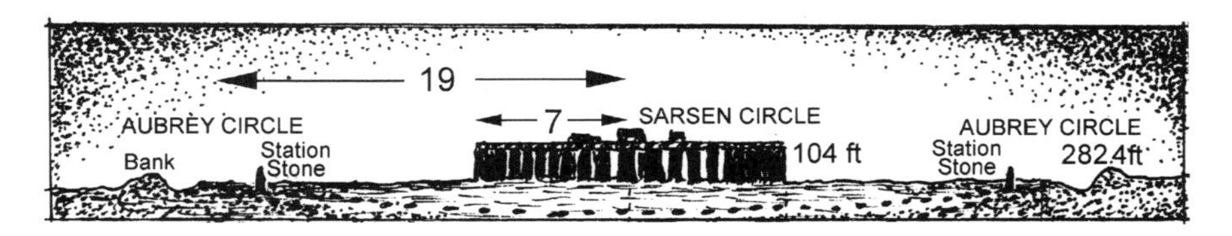

Figure 4.3 The two main circles at Stonehenge enshrine the builder's knowledge concerning soli-lunar cycles. The diameter of the Aubrey circle forms a ratio with the outer diameter of the Sarsen circle of almost exactly $^{19}/_{7}$ feet (*to 99.9%*). The reciprocal ratio is 0.368, the vital 'silver fraction' required for calendar and eclipse prediction.

The Stonehenge-Lundy Triangle

In 1992, while living near Carnac in Brittany, France, I enjoyed a further revelation. One afternoon, 'in a flash', I saw that Stonehenge itself formed the apex of a gigantic lunation triangle, with Lundy Island forming the right angle, *exactly* west of Stonehenge *(see figure 4.4)*. The bluestone quarry in the Preseli Mountains of West Wales defined the remaining point, while Caldey Island, near Tenby, even suggested a 3:2 point.

As Stonehenge is the only one of these sites built by human hands - the other three points on this triangle are all natural landscape features - three questions were immediately raised:

Was Lundy the starting point in this geomantic venture or was Stonehenge, the location of Lundy then being a remarkable coincidence?

Was Stonehenge sited where it is in order to complete a geomantic statement about the calendar across the British landscape?

Does this triangle provide the reason as to why the bluestone site was so important as a source of stone for Stonehenge?

Ever since this triangle was revealed to me, research has reinforced my belief that the answer to all three questions is yes. Stonehenge was most likely the *final* stage in a manifestation of astronomical, geodetic and geomantic wisdom, which predated the original siting of the monument, currently thought to be dated about 3,100 BC.

The implications of this took time to sink in. In 1993, I wrote up my findings in a book, *A Key to Stonehenge*. Calculations showed that this huge triangle was 2,500 times larger than that defined by the station stone rectangle, using Thom's megalithic yard as the basic unit. The right angle on Lundy was positioned almost in the sea just off Jenny's cove, which did not seem an optimum siting, and I was concerned that my calculations were not quite right!

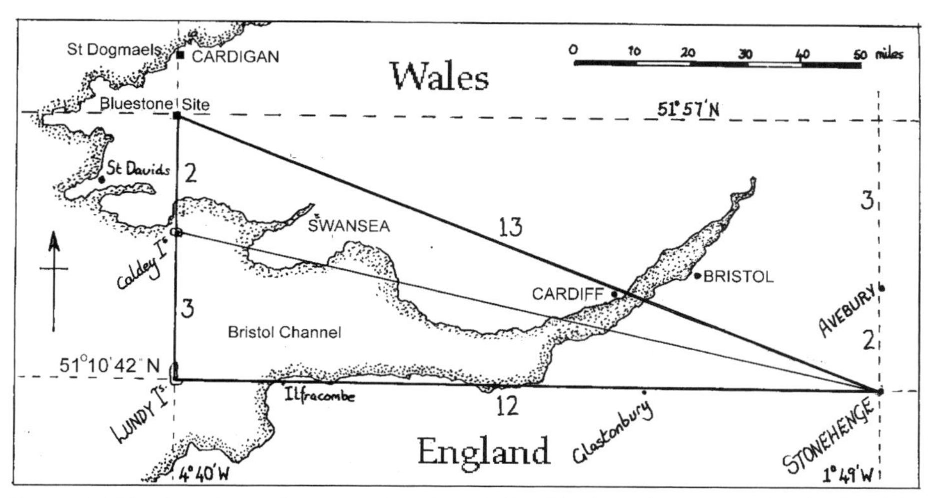

Figure 4.4 The Stonehenge lunation triangle, as defined by Stonehenge, Lundy Island, Caldey Island and the bluestone site. A geomantic symbol of soli-lunar wisdom from 3000 BC. The rectangle depicted above is 2,500 times larger than the station stone rectangle at Stonehenge.

A Metrological Cargo

There were more insights to follow. If a lunation triangle is made up using exactly one megalithic yard as the dividing unit, the 0.368 'over-run' turned out to be almost one English foot in length. Outside of the triangle perimeter lay a length of rope which was 99.84% of one Egyptian royal cubit in length, another astonishment. (*see figure 4.1*). So, a foot plus a royal cubit were to all intents identical to one megalithic yard. Here were three of the most ancient measures known to man related within a right angled triangle based on the astronomy of the lunation!

But all of these things were 'near-misses', very close approximations which were irritatingly not quite 100%. To make everything fit, all I had to do was reduce the value of Thom's megalithic yard, from 2.72 to 2.715 feet. This anomaly was to haunt me for many years. What justification there was to reduce the length of Thom's value for the megalithic yard?

At this point I decided to study ancient metrology, a decision that led me to re-read the works of John Michell, that beacon of geomantic and metrological research and the genuine father of the earth mysteries movement in post-war Britain. I had been much influenced previously by John's work. As a student in the 60s, both *The Flying Saucer Vision* and *The View Over Atlantis* had impressed upon me that modern science did not hold all the cards in the game of life upon the earth. Could John's later works shed light on the matter of my mismatch?

Following correspondence with John, I bought a copy of his *Ancient Metrology* whence I saw that this much ignored subject held enormous importance for the understanding of the prehistoric mind. Still, to my trained engineer's brain, the mathematics and the practical challenges in laying out this triangle seemed too complex for Stone Age Brits, and too unlikely. However, John's analysis of the metrology of Stonehenge, particularly his work on the dimensions of the Sarsen lintel circle, provided irrefutable evidence of an integer numerical relationship between that circle's inner and outer diameter, and the length and width of each lintel, all in terms of the six millionth part of the Earth's polar radius - a length of 3.4757485 feet, which Michell calls a royal yard.* I was becoming more and more intrigued.

Thus began a six year period where my material and John's pioneering work just would not quite mesh. Further complicating matters, Thom's megalithic yard apparently did not quite fit within the canon of ancient metrology either. In retrospect I failed to understand that the ancient measures were exact, and that their frequently long strings of decimals were in fact simple fractions. For example, a length of 1.07142857.. feet is simply 15/14 feet. Because of this mismatch of understanding, John declined to write the forward to my *Sun, Moon & Stonehenge*, published in 1998. I now understand his reasons - I had yet to understand the subtlety of the canonical system of Earth measures John had revealed. He in turn had not yet seen the illumination to be found from the geometry and astronomy of the lunation triangle.

Four years passed and still the numbers would not mesh. Despite correspondence and the occasional meeting, John and I could not find a way to marry the megalithic yard and the lunation triangle into his established numerical system for ancient measures. It took a third party, the tenacious metrologist John Neal, to provide the catalyst. Neal's *All Done with Mirrors*, the most rigorous exposition on ancient metrology to date, entered my life as would a bucket of ice-cold water during sleep. Having extended the range of units defined in Michell's *Ancient Metrology*, Neal brought to light an integrated system, connecting most of the known units of measure in the ancient world. This system was based on the English foot, that same foot that had spilled out from my astronomical work on the lunation triangle.

By a fortunate twist of fate, Neal had strong links with the area of West Wales where I live, and he became a regular visitor. Without these visits it is unlikely that Michell and I would ever have resolved the difficulties which had separated our work. Neal challenged me to prove the megalithic yard as a valid unit of ancient metrology. I explained to him how the soli-lunar astronomy of

* The royal yard is 2 Egyptian royal cubits at 'geographic' value, 1.737874285... feet *(see figure 2.9 and pages 81 - 87)*

our planet defines the ratios between three most ancient measures known. If the astronomical megalithic yard represents the lunation period, then the foot represents the difference between the solar and lunar year and the royal cubit then defines the eclipse year. But ratios and time periods are not actual lengths. Someone, sometime in prehistory, decided on actual lengths for these ratios, lengths which are related to the key dimensions of the earth.

The Ubiquitous English Foot

In addition to the foot emerging from my work on the lunation triangle, it crops up elsewhere, within a geodetic context. A recap may be helpful at this point. The ancient geodetic value for the equatorial circumference of the Earth is 24,902.94857.. miles, which is 131,487,568.5 feet, or 365.24325 x 360,000 feet. Such a factoring is exactly how astronomers and scientists, now as in ancient times, would wish to calibrate that great flywheel in space, our planet, and thereby reconcile time and space, angles turned with distances travelled. The fact that the foot is enshrined within this factorisation of two fundamental measures - of time and angle - suggests that the foot is an astonishingly enduring artefact bequeathed to us from the prehistoric world. When the ratios emerging from the lunation triangle, rounded up as 1 : 1.72 :2.72, define the soli-lunar rhythms, that is a remarkable thing. But when, as lengths in feet (and not rounded up!), they become familiar prehistoric measures exactly relating to the size of the earth, that is quite another thing, and demonstrates the existence of a highly evolved prehistoric science.

The Initiation into Ancient Metrology

Many people upholster their brains with the comfortable notion that the ancients could not have been capable of high accuracy in their work. Neal points out that the taking of *average* lengths for the multiple examples of feet, cubits, rods, yards to be found in museums is almost solely responsible for this erroneous belief. Each discrete measure is exact, expressing a simple fraction of the foot or mile, as amply demonstrated in chapter two. Averaging them only obfuscates the system. Ancient metrology is a coherent and wholly accurate system of measures, based on pure number ratios. The subject appears complex, yet in fact is awesomely simple. It is also awesome in its implications for historians.

To understand this ancient system required a suspension of some prejudices inherent in my modern engineer's mind. The ancient system was then revealed, and I have Michell and Neal to thank for that. Mathematically, it required of me only a knowledge of ratio and proportion, a pocket calculator with memory facilities and time playing with the known lengths bequeathed to us from the ancient world. The system reveals itself in all its glory. An aspirant need only acquire a copy of *Ancient Metrology, All Done with Mirrors* and this book. You may be some time, and you must be prepared to wear out a couple of calculators!

The Emergence of Common Ground

In April 2002, I visited John Michell and John Neal in London. It was a good night. At the time John Michell was hard at work on alignments connected with Stonehenge and the latitude degree between 51° and 52° north. I drew his attention to the string of islands and sacred/holy sites which projected from the shortest side of the lunation triangle. These all lie near to longitude 4° 40′ west - Lundy, Caldey, St Dogmaels Abbey, Bardsey Island and Castell Odo, Holy Island on Anglesey, the Calf of Man and so on up to Cape Wrath, the northwestern-most tip of Scotland. I had also noticed that the meridian line from Stonehenge passed up through Holy Island near Lindesfarne, and had termed these two parallel lines 'Jacob's Ladder', thinking them to be important in understanding prehistoric surveying.

John responded warmly to this, having also recognised and studied the Stonehenge - Holy Island alignment. He asked what my estimate was for the length of the '12' side in the giant lunation triangle, and I gave him my easily remembered figure of 123.4 miles. That was the crucial turning point, for Michell immediately recognised the basic unit of measure in the triangle and its place within the system of ancient metrology, described by him in Book Two. This realization widened the perspectives of us both and was quickly to reveal the true length and metrological importance of the megalithic yard. The subject we were each investigating suddenly became larger. We decided to work together on describing it. That was the genesis of this book.

In September 2002 we took a working holiday on Lundy Island, to inspect the calculated point of the right angle within the large lunation triangle. The point was readily found using a GPS device, although even without it the relevant place is clearly identifiable, being a large 'tump' half way along the north-south axis of the island and located at precisely the same latitude as Stonehenge.

The Lundy discoveries are described next as part of an investigation undertaken by John and myself of all four sites defined by the lunation triangle; Lundy Island, Caldey Island, Stonehenge and the Preseli Mountains of West Wales. The reader is invited to share this voyage of discovery.

Chapter Five

A Visit to Lundy and West Wales

The Preseli Triangle

> I will show you an island. Come with me
> to where, in the west, Wales falls into the sea
> and washes her hands of history.
>
> Dr Raymond Garlick

What a place is Lundy Island! Located in the middle of the Bristol Channel, equidistant from Devon and South Wales, Lundy is a three mile long and half mile wide wonder, a high plateau raised up out of the ocean on a skirt of sheer cliffs. Long famous for its wildlife and for providing some of the most challenging ascents for rock-climbers, Lundy has hardly been affected by the twentieth century, let alone the twenty-first, and sits majestically on its cliff throne oblivious of the hectic pace of life on the British mainland.

Lundy presently takes its name from the Scandinavian word for a puffin, and this charming, parrotty sea-bird is still to be found here, though no longer abundantly. In Welsh, the name for the island is *Ynys Elen*, which means 'The island of Elen' or 'The island of the elbow, bend or right-angle.' The modern church on the island is dedicated to St Helena, the Christianised derivative of the Celtic Elen, as was the original church which now lies in ruins, surrounded by its graveyard, at the highest point on the island, on Beacon Hill, adjacent to the Old Light. This single historical fact - the Welsh name for the island - provided initial supporting evidence for the Stonehenge-Lundy lunation triangle, whose right angle* lies right across the middle of the island.

* The letter 'L', pronounced 'el', is itself a right angle.

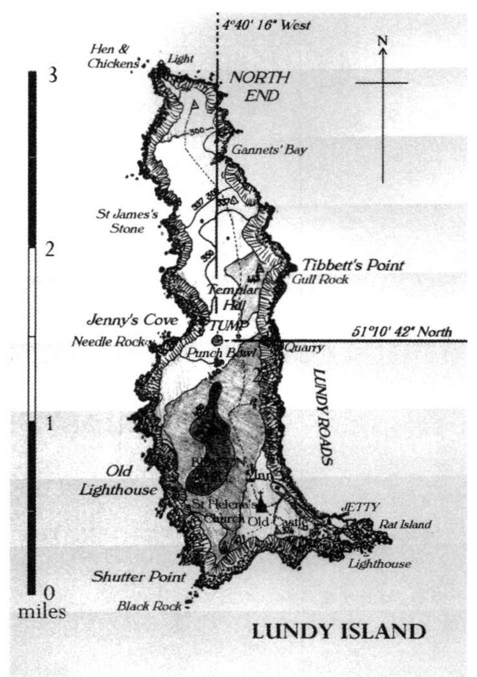

Fig 5.1 Lundy Island, showing the features mentioned in the text. The island is 3 miles in length, the site of the right angle *(marked above)* is located at the centre of the island

Lundy has a long history, but much of it remains shrouded in mystery. In the Bronze and Iron Ages, the island was used as a stepping stone between the coastlines of Wales and Cornwall. By 450 BC, Carthaginian traders had links with the silver miners of the Bristol channel coast, and by 150 BC the channel had become an international trade route. Ptolemy compared the passage between Hartland and Lundy with that of the Straights of Gibraltar, and Hartland he named Heraclis Promontary, after the Pillars of Hercules. No evidence of Romanization has been found on the island, although there is a Roman reference to its occupants as 'a specially holy race of men...who refused trade and had visions of the future'. Wise men.

The importance of Lundy within Celtic mythology is emphasised by historian Geoffrey Ashe, who writes, 'In Welsh it has an uncanny character and is supposed to be *Annwn* or rather a place where mysterious realms can be entered. Various hills and islands are points of access to it. The most important hill is the Tor at Glastonbury, the most important island is Lundy... *Annwn*'s inhabitants have human form but are not strictly human. They are immortals - fairyfolk or demons according to one's point of view. Some are Gods thinly disguised. Living humans can enter *Annwn*, and so are spirits of the dead, but it is neither a heaven or hell in the Christian sense. To a certain extent it resembles Avalon...'

Lundy was thus regarded as an 'Isle of the Dead' to the Celts, who regarded it as a holy island, an Otherworld in the west to which their illustrious heroes were ferried to join with their ancestors, the place of the setting sun and therefore a fitting place for a solar hero to be buried. If Lundy was indeed Avalon, the mythology of Albion's greatest solar hero, the fifth century King Arthur, would certainly fit the geographic placement of Lundy, lying as it does almost directly west of Glastonbury, our modern Avalon. And Elen presided over both the ancient straight trackways of Britain and also the west, the direction of the setting sun. The notorious Giants' Graves discovered on Lundy in 1856 were alleged to have revealed skeletons 8 foot in length. Despite recent archaeological evidence that the graves date from the fourteenth century, this discovery upheld the traditional mythology of Lundy.

Contemporary with the arising of the story of Arthur, the Early Christian church in Wales reached its greatest flowering, with St Illtud's college at Llantwit Major, and monasteries at St David's and St Dunstan's, all equidistant from Lundy *(see figure 1.2)*. A wealth of saints, Sampson, Govan, Madoc, Patrick, Gildas, Ishmael, and even Elvis embellish the landscape within sight of Lundy Island, as church names, town names and parish names.

Lundy appears historically to have had more connection with Wales than with Devon. Because the island fails to get a mention in the Domesday book, a Norman audit of British assets undertaken in 1086, it is clear that the island had not yet passed into Norman hands, as was also the case for much of West Wales. Furthermore, the island is named after a Welsh saint, and in the later Pipe Rolls its name is recorded as Ely, '*of Elen*'. The first Normans to take title to the island were the De Newmarchs, the Marisco family subsequently renting the island for 'a fifth part of a knight's fee', in 1166.

Almost all prehistoric traces on the island have been obliterated by subsequent changes. Whilst the desolate north of the island retains its ruined hut

Fig 5.2. Recycled megaliths in the corner of the graveyard at the site of the original church, dedicated to St Elen. The left hand stone was dedicated to Potinus, possibly St Patrick's grandfather; the middle stone, once much taller, to the son of Vortigern, an important British chieftain. The location is Beacon Hill, adjacent to the Old Light. *(photograph by Paul Broadhurst)*

circles, tantalisingly suggesting a busy community living here during the Bronze Age, only the area around the old churchyard retains significantly large standing stones. Elsewhere, the paucity of good building stone to be had on the island has meant that it all gets recycled to construct the austere granite buildings which now stand firm against the Atlantic gales. Apart from an odd classical mansion, ruined hospital and the humble dwellings of farm workers and visitors, the only buildings of any stature on the island are the Old Light, adjacent to the site of the original Church, the Castle Keep, on the extreme south of the island, the modern stark and austere church of St Helena, and the curious house near the 'quarter wall' at Tibbett's Point.

To those seeking the amusements of the twenty-first century, Lundy must appear a bleak and forebidding place where nothing much changes and where there is nothing much to do. In truth, the island presents one of the few remaining places in the British Isles where one may retreat from the babble and complexity of modern life. Within its windswept and rugged beauty may be found a glimpse of Old Albion, a wild place where one feels that the ancestors may be readily contacted. To the north one can clearly observe Caldey Island and the Preseli Mountains, home of the bluestone outcrop. All three sites are intervisible.

Modern day boats bring visitors from both sides of the Bristol Channel, from Swansea in South Wales and from Ilfracombe or Bideford, depending on the tide, in North Devon. The jetty at Rat Island disgorges hundreds of eager passengers many of whom become considerably less eager at the prospect of ascending up a steep track for nearly a mile to the tiny village which is the modern heart of activity on the island. Breathless visitors are rewarded with the prospect of an excellent meal and a pint of 'Lundy Experience' or 'Lundy Light' in the Marisco Tavern. Alternatively, there's the kind of shop once common throughout Britain where they stock one of everything and two of most, alongside the inevitable souvenirs, postcards and the unique Lundy stamps, available in various demominations of the island's currency, the Puffin.

Day trippers have just a few hours to ramble across this ancient landscape before they must descend back down the steep pathway to Rat Island, the jetty and a return to the twenty-first century. Those who book to stay on the island, or live and work here, watch quietly and heave a sigh of relief as the noisy hoards disappear and leave the island once more to its few inhabitants. From three o'clock onwards, a tangible peace returns to the place as the visitors depart.

The night life centres on the Marisco Tavern. Freed from the cacophany of fruit machines, juke box and perpetual mobile phone jingles, the atmosphere

Fig. 5.3. The austere beauty of the Old Light, from an 1820 elevation. Designed in solid granite with a stone roof, this classical lighthouse stands firm on Beacon Hill, the highest point on Lundy. It stands near the ruins of the old church and graveyard, replete with standing stones.

probably has not changed much since Nelson's time. Outside the tavern, with no street lights to ruin the night sky, the visitor wends his long way back to the Old Light or Castle Keep, a journey which connects in the most forceful way possible to the importance of the waxing and waning moon in her cycle.

Near to the full moon, the journey is an easy and silvered experience of grey light and shadow, while near to the new moon one stumbles haplessly from field boundary to field boundary, randomly encountering thistles, nettles and potholes. Listening to visitors undertaking this trek, one could be forgiven in thinking that such experiences spawned the origin of swearing.

In September 2002, it became possible to take up a booking cancellation which offered nine places for five nights, in the Old Light and the Castle Keep. Neither John nor I had ever been to the island, and for ten years I had been promising myself that I must go and see just where the right-angle of the lunation triangle was located, but, as such things happen, never did. Earlier in the year, John, metrologist John Neal and earth-mysteries author Paul Broadhurst had eagerly suggested we make such a trip, so all nine places were booked, the remaining spaces being filled up with family and friends.

It was to be a magical holiday blessed with the best Indian summer for many years. Five days of sunshine is a rarity here, and the island community was in the grip of the worst water shortage for many years. Burdened with compass, theodolite, GPS, cameras and all the paraphernalia of surveying, I was greatly relieved to discover a Landrover available to lug heavy items up the long ramp to the Old Light, which was to be our home. John Neal, who is blessed with a talent for such things, scammed a ride up, whilst the rest of our group walked eagerly up, weaving between the boatload of nearly two hundred grumbling, panting and perspiring day trippers who had not figured hill-climbing to have been on the *menu de jour*.

John and I quickly made an inspection of the location identified as the great right angle in the lunation triangle geomancy. John was quick to point out that, from its northmost to southernmost tip, Lundy is three miles in length. Almost exactly in the middle of the island is a large flash pond, fed by several freshwater springs and rimmed by a retaining causeway along its western boundary. The area is marked 'the Punchbowl' on maps. The pond covers about two acres, and beneath the wall to the west is the inevitable marshy ground fed from seepage from the pond.

The area immediately around this flash pond quickly established itself as being the most enchanting place on the island. A picnic was planned for the following day, duly held on a soft grassy ledge just under a rock formation appropriately called 'the cheeses'*(opposite)*. This impressive natural stone temple faces west to the Atlantic, and overlooks Needle Rock, which very easily can be made to look like an enthroned Queen Victoria, or a Mermaid coiled up with a dragon, depending on the light and the imagination of the viewer.

Fig 5.4 The Tump that marks the centre of Lundy and also the right angle of the Stonehenge-Preseli triangle. The tump, flash pond and right angle can be located on the map on page 46. Although marked with an arrow, the change in vegetation clearly identifies the tump, whose top contains a peppering of small prehistoric stones in addition to a modern field boundary wall.

Elin - The Right Angle

elin n. angle, bend or elbow
Spurrell's Welsh Dictionary

North of the flash pond is a distinct mound or 'tump' (*marked above*), the flat summit of which contains various ancient large stones and a stone field boundary. Beside the lichen covered stones of this ruined wall I discovered John investigating a curious and completely unexpected modern feature, a well made stone and slate letter box for the sole use of members of the unlikely *Lundy, Centre of the Island, Last of the Summer Wine Club*. The box contained an impressive rubber stamp and a commentary book which contained spirited exchanges about life, the universe and everything. A *pythonesque* sense of the British spirit emerged as we scanned through these messages, written by the known and the anonymous, the inspired and the downright vulgar, at the centre of Lundy. We added our geodetic contribution.

Fig 5.5 The Cheeses. Situated at the same latitude as Stonehenge, this natural stone temple faces the Atlantic, some two hundred yards west of the 'tump'.

And this proved to be the place of the right-angle, the central point, the highest point in the middle of the island, where the east-west line from Stonehenge elbows its way northwards to the bluestones of Preseli, via Caldey Island. Our quest to identify the site was done, although the presence of this central place had already been recognised by the founders of the above mentioned club. All that remained was for Neal, Michell, my brother and myself to finally agree on terms and units and, in the remarkably conducive atmosphere of the Marisco Tavern, a breakthrough occured.

The Drusian Foot

John Neal had clearly been giving the matter a great deal of thought. The whole business of lunation triangles and astronomical megalithic yards was throwing a curved ball at metrology and Neal wasn't accepting of my smaller megalithic yard without thorough scrutiny. However, at one pint in the evening, calculators were drawn, and he leaned across the table and began talking about the Drusian foot, a clearly defined ancient unit of length attributed to Nero Claudius Drusus, and widely in use in Northern Europe during the Roman Empire. It is a measure of 1.08617143 feet. The Drusian foot connects to the other values of the foot through the ancient metrological system. It is a 'geographic' measure whose root foot is $^{15}/_{14}$ths of the English foot *(see figures 2.8 and 2.9)*.

Within a few minutes we established that a unit of two Drusian feet defines the station stone rectangle at Stonehenge, with its diagonal, as 50:120:130, decimally matching the 5:12:13 proportion. The short sides become 100 Drusian feet in length. Two and a half Drusian feet are then exactly equal to my 'astronomic' megalithic yard of 2.71542857 feet. Neal suggested that the AMY might better be described as a megalithic 'step', as metrologically it is a 'half-pace' related by the fraction $^{5}/_{2}$ to its parent foot, rather than the division by three always associated with yards. *(This is confirmed on page 14, where the half-pace or step identified from Wright's treatise on navigation, and based on the Roman foot, is exactly $^{24}/_{25}$ of one AMY)*. Finally, if the Drusian foot is multiplied by $^{24}/_{15}$, the geographic royal cubit (1.7378425 feet) is revealed, two of which define the royal yard, used in the design of the Stonehenge lintel structure.

These fractional relationships *(see figure 3.3)* show the pedigree of the astronomical megalithic yard within the system of ancient metrology, the Drusian foot resolving my previous anomalies with the length of the megalithic yard. The astronomy, the stone circles and the metrology were finally married together in precise figures during that evening on Lundy, and we toasted the occasion.

Four researchers, who had previously each been clutching separate pieces of a jigsaw, were now able to see how their pieces fitted together. We had formed a new and clear picture of the working techniques used by our prehistoric forebears. Our terminology, numbers and artefacts were in complete agreement, a seamless garment, the proof of which was being displayed right across all the decimal places of our calculators. The lunation triangle had acted as a catalyst for this entire process, appropriately undertaken on Lundy.

Five glorious sunny days on Lundy had closed a chapter in our work together and opened a huge door into the prehistoric world. Lundy was kind to us, and we were all blessed by the Lundy experience.

A Visit to Caldey Island
and the 3:2 point

Caldey island is another jewel. Lying just off the South Pembrokeshire coast, near to the popular resort of Tenby, Caldey has an ancient association with the early Christian church, and many Celtic saints are associated with the Island. The original church remains dedicated to St Illtud, and Samson and Gildas both featured in the establishment of the spiritual community there during the 5th century AD. The present order was founded by Pyro in 535, whence the island received its Welsh name of *Ynys Byr.*

Fig 5.6 A geological map of Caldey island, dated about 1930, with relevant sites marked. The island lies about a mile from the seaside resort of Tenby, in Pembrokeshire.

In the 12th century, Benedictine monks from St Dogmaels abbey, on the estuary of the river Teifi, near Cardigan, established a priory on Caldey. There they remained until the Dissolution in 1536. The medieval priory survived and, in 1906 Anglican Benedictines bought the island and built the Italianate style abbey which now stands proudly above the tiny village. Financial problems forced the order to sell Caldey in 1925, eventually passing on the ownership to the stricter, more contemplative, Cistercian order in 1929. These new monks originally came from Belgium, and re-established the long tradition of Cistercian monastries in Wales - there were thirteen such monastries in 1536. Presently, the monastery thrives, due in part to increased tourism and the manufacture of a unique range of perfumes by the monks.

The island is thirty miles south of the abbey at St Dogmaels, and has all the qualities and atmosphere required of a natural sanctuary. As is the case with Lundy, the island issues its own postage stamps, with the island's own unit of currency, the Dab, named after that small flatfish which may be caught from the beaches here. Writing postcards while taking afternoon tea at the colonial style tea-shop on the Green at Caldey is a rare and *olde worlde* experience, and a day trip to the island may be heartily recommended.

Fig 5.7 Pence, Puffins and Dabs - The four points of the lunation triangle expressed through the stamps of three currencies, two countries and two islands.

The geodetic line passes over the eastern corner of the island. Maps of Caldey indicate an obelisk here, about half a mile from the site of the original priory. If once a prehistoric site, there is today no evidence of prehistoric origins. The 3:2 point lies less than half a mile off-shore to the south of the island, an undersea hazard called Drift Rocks on the old maps. Caldey lies within 99% of the ideal location required to fulfil its geomantic role in the lunation triangle, a close but ultimately unsatisfactory conclusion to a geodetic exploration which otherwise proves entirely accurate.

Yet Lundy, Caldey and the Preseli site are all natural landscape features, points of high ground appropriately arranged for the purposes of surveying along a north-south line. It would be astonishing if Caldey were to fall precisely on the 3:2 point, and 99% is surprisingly close. Caldey's position in no way affects the precision by which the large 5:12 rectangle was laid out and from which a 2500:1 replica was later constructed at Stonehenge, around 2500 BC.

Modern visitors to Caldey must take a small boat from nearby Tenby. In fine weather, the half-hour trip is a delight, and Caldey offers a wealth of fine secluded beaches and abundant wildlife in addition to its historical riches. St Illtud's church is reputed to be the oldest Celtic church in Wales *(opposite)*. With its stone spire resembling a megalithic standing stone and its magnificent cobbled floor, this church holds an atmosphere quite unlike modern churches.

Nearby, in the ancient monastery grounds, may be found a circular well, overgrown and neglected, in the centre of which has been placed a large square-section megalith. This coming together of prime male and female symbols is a pagan arrangement unlikely to be other than an embarrassment to the present order, perhaps one reason for the neglected state of these ancient gardens. The more esoteric numerological meaning of the numbers 2 and 3, espoused in all the traditions from the Rig Veda through to Pythagoras and later texts, offers some confirmation that the geodetic meaning was understood, for where else may one find a megalith sited within a circular holy well?

In other books I have shown how the geometry and numerical basis of the lunation triangle equates with the Pythagorean tradition, as a symbol of the *Sacred Marriage,* expressed as Sun and Moon integration within the calendar. The 3:2 point on the triangle places the first 'male' and 'female' numbers together, joining them and creating a harmonious musical fifth, solving the problems of the calendar. The square of the intermediate hypotenuse is 153, and whether we call this integer 'fishes in the net' as St John does (Chapter 21), or mathematically as its square root, 12.369, the link between Christianity, Pythagorean teachings and megalithic astronomy is made abundantly clear.

Fig 5.8 St Illtud's church. A unique megalithic spire, cobbled floor, and attached medieval priory and gardens gives the site an astonishing atmosphere. Possibly the oldest Celtic church site in Wales.

As for so many sacred sites, St Illtud's church and priory is sited on an underground water source, which once nourished the whole garden complex and without which human habitation would have been rendered much more difficult. The presence of underground water at the two most important ancient and prehistoric sites on Caldey and Lundy, both hill-top sites, suggests why these locations were chosen for settlement. When this choice was first made remains unclear, although mesolithic and prehistoric remains have been discovered in large numbers on both sites.

In *Secret Shrines,* author Paul Broadhurst comments that, 'Many of the old hills, often now with churches perched on top, possess natural springs which are the physical embodiment of the female force that balances the (male) dragon

Fig 5.9 An eight foot tall square-section standing stone sits centrally in the circular well within the overgrown gardens of St Illtud's priory, symbolic of male and female energies.

power'. Nowhere is this better realised than on Caldey Island, where the symbols of paganism and Celtic Christianity meld in harmony, supporting the larger astronomical and geodetic symbols contained within the lunation triangle.

St Illtud was, according to Gildas, 'the polished teacher of almost the whole of Britain'. His main teaching college and monastery at Llantwit Major lies adjacent to the intermediate hypotenuse of the lunation triangle and it was he that established the priory and church on Caldey Island in the sixth century.

The Top of the Triangle

Gors Fawr Stone Circle and Carn Wen

Travelling due north of Caldey the '5' side of the triangle meets the Pembrokeshire coast between Monkstone and Amroth, the terminus for the world famous Pembrokeshire Coastal Path which begins in St Dogmaels. The alignment courses up through Pembrokeshire, the two-mile wide band from 4°40′ to 4°43′ west embracing in its travels a longstone, several standing stones, prehistoric earthworks, burial chambers and cairns and, eventually, the Glandy Cross Prehistoric Complex, which includes the mountain Carn Wen.

Of note here are Meini Gwyr burial chamber, Gors Fawr stone circle, the Preseli bluestone site, Carn Wen, Carn Besi, Bedd Arthur (Arthur's grave) and Carn Arthur. The band continues past the triple cairned mountain peak of Foel Drygarn, to St Dogmael's Abbey and Cardigan Island, again including many prehistoric sites. There is no greater concentration of megalithic sites than to be found within this narrow strip enclosing the '5' side of the lunation triangle.

Five roads converge to define the modern village of Glandy Cross, on the A478 in North Pembrokeshire. Here may be found the remains of a large and once important Neolithic and Bronze Age site, known as the Glandy Cross Prehistoric Complex. Archaeologist N P Figgis, in her recent book *Prehistoric Preseli,* notes that it 'is the least known and most important gathering of 'ritual' sites in West Wales, and one of the most significant along the whole length of the British west coast.' It is not hard to discover why she should hold such an opinion. Within a few miles of Glandy Cross may be found the evidence for many ritual sites, a large linear cemetary, two axe making factories, a possible cove, kerbed cairns, stone circles, single and double standing stones and, of course, Carn Meini, the site of the Stonehenge bluestone quarry.

In 2002, the author, accompanied by his brother, made a theodolite survey of one ruined ring-cairn just north of Glandy Cross. The geometry proved identical to Thom's 'Type B' flattened circle. The centre of the ring was unusually defined by a stone, and the axis of symmetry of the ring was orientated directly north-south. This design is routinely found at significant sites, it forms one signature of the megalithic circle-builders art, the geometrical and astronomical significance of which may be explored in a previous book, *Sun, Moon & Earth*.

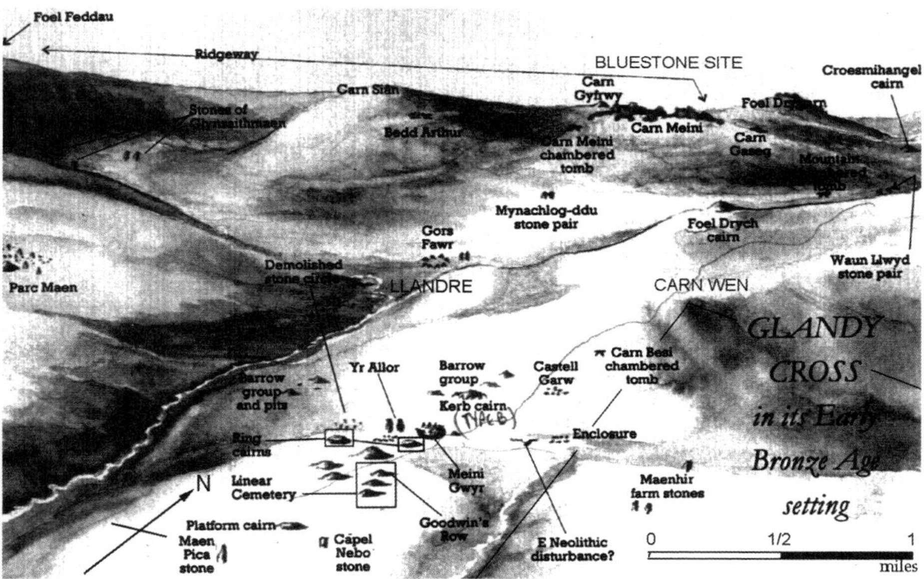

Figure 5.10. The Glandy Cross Prehistoric Complex. According to archaeologist N P Figgis, 'the least known and most important gathering of 'ritual' sites in West Wales, and one of the most significant along the whole length of the British west coast'. The locations of the bluestone site, Carn Wen, Llandre, compass directions and the outline of the lunation triangle have been added by the present author. *(Illustration from Prehistoric Preseli, by N P Figgis, published by Atelier Productions, and reproduced by kind permission of the author.)*

The top of the Stonehenge Lunation triangle is a distance of 100,000 AMY directly northward from the central point of the 'tump' on Lundy. The calculations to locate the point are shown in appendix two. The location is the summit of a hill called Carn Wen, Welsh for 'white cairn'. Today there is neither evidence of the cairn nor its white quartz covering from which, presumably, the hill got its name. There are some impressively large quartz boulders at the

Fig 5.11 Carn Wen - 'White Cairn', the top of the Stonehenge-Preseli Lunation triangle. This uniquely flat topped mountain lies within the Glandy Cross complex *(see previous page)*. Any white cairn has long since disappeared.

foot of the hill and local churches have their church walls topped with tons of the stuff. During the past century much of the north side of the hill has been quarried, although an Ordnance Survey 'trig' point still resides on its summit. An arduous climb to the summit awaits the casual visitor, with uneven ground and much gorse.

Unique to Carn Wen, its summit is flat and level and of such an area that it would be possible to site two football pitches there. On the earliest Ordnance Survey map of the area, the hill is clearly shown with its flat top, and stepped sides are drawn to the south and east. The view from the summit is awesome, encompassing the Gower peninsula, the Brecon Beacons and, to the south and west, the entire Glandy Cross Complex. To the north, there is a good view through the only natural gap in the Preseli range.

The summit of Carn Wen provides an ideal natural location to define the apex of the lunation triangle. Whether its uniquely flat top is natural or man-made will be for others to decide, as also will be the matter of what happened to the white cairn, from which the hill takes its name and of which no trace can be seen today. There are some large bluestones to be found in walls and in scree lower down the hill and the bluestone site, Carn Meini, is strongly and prominantly visible from Carn Besi, just down the modern A478 road from Carn Wen. The tragically ruined site of Meini Gwyr *(below)*, once an important focal point for the Neolithic inhabitants of Glandy Cross, is less than a mile distant.

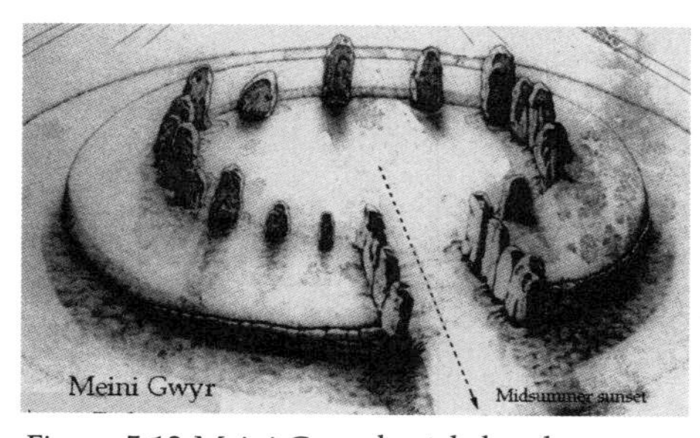

Figure 5.12 Meini Gwyr burial chamber as was, from an artists impression. Two stones remain and the concrete plugs that once identified the placement of the other stones have been removed and left at the nearby field boundary.

A Fork in The Road

At this juncture we must investigate two possible routes by which the huge lunation triangle may have been defined. If the surveyors had been using true north-south meridians to define the triangle, then its form would be a spherical triangle, and though the right angle on Lundy would be ninety degrees, the '5' side would, in its journey northwards, progressively bend slightly inwards as the two-dimensional triangle was laid down onto a three dimensional planet. If however, the surveyors were using plane chart surveying, assuming the earth's surface to be flat for the purposes of their grand vision, then the '5' side from Lundy would have to track parallel to the true meridian at Stonehenge, splaying outwards slightly as it progressed northwards. Either technique would introduce an inevitable distortion.

Figure 5.13 Gors Fawr stone circle. Lying under the bluestone site, the circle contains sixteen stones plus two large outliers, which indicate the midsummer sunrise position.

This problem is well known to all map-makers, and is somewhat akin to that posed when attempting to paste wallpaper onto an uneven or curved wall. Two-dimensional maps define the distortion through a clearly defined projection. For our purposes, if the triangle was laid out using true meridians, then its apex sits on Carn Wen, while if plane chart surveying was applied, using parallel 'meridians', then the short side of the triangle passes from Lundy northward through the only tumulus on Caldey, thence to the ruined monastery of Mynachlog ddu ('black monastery'). At the exact location (51° 55′15″N; 4° 43′ 35″ W) is a large bluestone megalith incorporated into the wall of a stone barn at Llandre Uchaf. Thereafter the line progresses northwards past the famous stone circle of Gors Fawr, with its large outliers aligned to the midsummer sunrise. The route up the defining 'ladder' is illustrated in appendix four.

Gors Fawr is the best known surviving stone circle in West Wales, and magnificently located. To the north of the circle, which resides on level ground amidst an almost 'lunar' landscape surrounded by the soft peaked and undulating Preseli mountains, lies the bluestone site itself, Carn Meini, a craggy outcrop which dominates the skyline *(illustrated in figure 5.10)*. The projected line runs past it to Carn Arthur and Bedd Arthur, Arthur's Grave, a curious boat-shaped assemblage of bluestones.

Both sites are equally valid as contenders for defining the top of the '5' side, the author favouring Carn Wen, whose flat summit platform would have been an ideal location to erect a white cairn to mark the top of the lunation triangle. Both sites lie within 2½ miles of the bluestone site and both fall within the boundaries of the Glandy Cross complex.

PREHISTORIC PRECISION

The integration of metrological, geometrical and geodetic design which defines Stonehenge and the Lunation Triangle

STAGE ONE

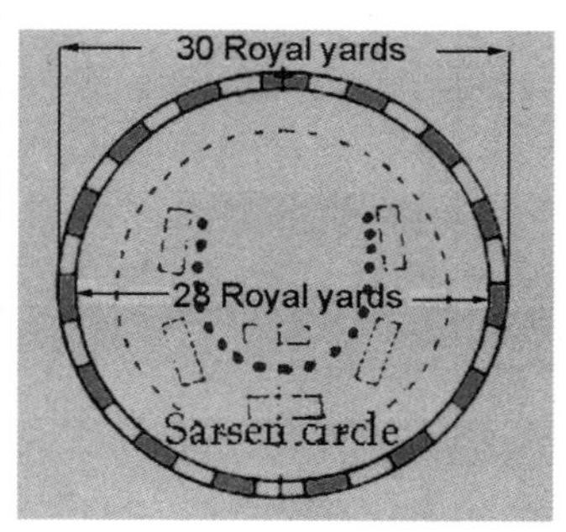

The outer diameter of the sarsen lintel circle is exactly $^{15}/_{14}$ths the inner diameter, as 30 to 28 royal yards.The royal yard is a six-millionth part of the polar radius.

If $^{15}/_{14}$ft is taken as the 'root' foot, the 'geographic' value becomes exactly $^{15}/_{14}$ths of the Greek foot of 1.01376 English feet *(see chapter two, figure 2.2 and 2.8)*. The inner sarsen lintel diameter is exactly 96 such units.

This unit was identified by Caesar Nero Claudius Drusus during the Roman occupation of northern Europe. The 'Drusian' foot is exactly $^{15}/_{48}$ths of the royal yard.

STAGE TWO

In units of two Drusian feet, the station stone rectangle at Stonehenge measures exactly 50 by 120, the diagonal 130, the metrology numerically matching the geometry.

The Astronomical Megalithic yard (AMY), as defined by the time periods of the lunation *(see Chapter Three)* is exactly $^{5}/_{2}$ Drusian feet.

In units of 8 AMY, the Station stone rectangle measures exactly 5 by 12. Its diagonal, also the diameter of the Aubrey circle, is therefore 13 of the same units, the metrology again matching the geometry.

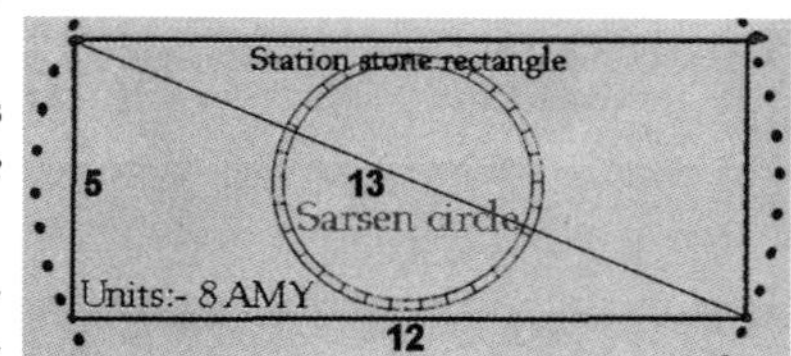

STAGE THREE

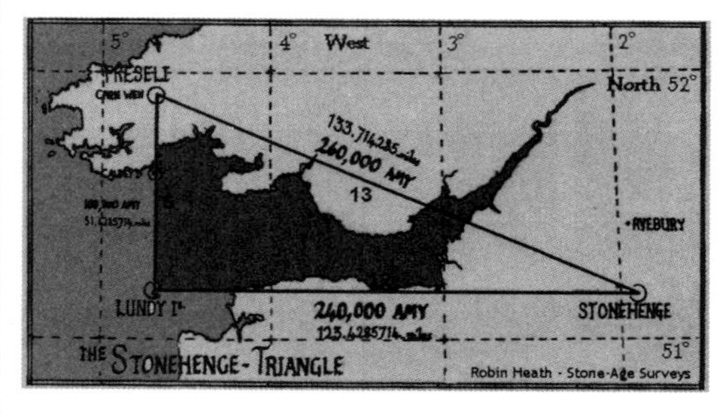

If the station stone rectangle is scaled up by 2,500 : 1, it exactly matches the Stonehenge-Preseli lunation triangle. In units of exactly 20,000 AMY, the dimensions are 5 : 12 : 13, the metrology again matching the geometry.

10,000 AMY = 36/7 miles, the unit identified by John Michell as that used in the prehistoric survey of Britain, revealed in Book Two.

Stonehenge was located in order to define the lunation triangle. The chronological sequence of events therefore runs in reverse order to that shown above, the erection of the sarsen circle being the final part of the process.

Chapter Six

Prehistoric Precision

A Summary of Book One

The most important questions that remain to be answered concerning Stonehenge are, "Why were the bluestones so important to the construction of Stonehenge?", "Why was Stonehenge built where it is?" and most importantly, "Why was Stonehenge built at all?"

The material presented here provides a robust answer to all three questions. In addition it places the megalithic yard firmly within the metrology of the ancient world and reveals its astronomical connection with the moon. Megalithic architecture still holds many secrets, but we now have a reason why Stonehenge was built where it is, where before this reason, and its meaning, was obscured. From 1923, when petrologist Dr H H Thomas determined once and for all time the source of the bluestones, it was again possible to read the geodetic and astronomic message from Stonehenge. That single piece of information unlocked a long-closed portal into the hall of *Annwn*, the Otherworld of our prehistoric past. Down through the ages, those smouldering Preseli Mountains guarded their secret well, yet after 1923, Stonehenge and the Preselis became unequivocally connected, a bond literally written in tablets of stone.

But could the lunation triangle be a fiction, a *chimera*? Well, no, because even if the existence of the Stonehenge triangle is denied, it remains a fact that Lundy, Stonehenge and the bluestone site form the three corners of an accurate 5:12 rectangle, one which is 2,500 times larger than the station stone rectangle. No, because the unit of length common to both is the astronomical megalithic yard defined in chapter three, based on the immutable calendrical realities of the sun, moon, earth system. No, because the megalithic yard so defined integrates perfectly within the known metrological framework of the ancient world. And no, because the construction is related to the size of the earth, so that the '12' side is exactly 1/32nd of the earth's polar radius while the '13' side is an astonishingly lunational 1/29.53ths!

In reviewing the evidence presented here, the reader might also consider the graphic on page 60. The 'geographic' Drusian foot, which defines both the Stonehenge rectangle and the large lunation triangle in multiples which are themselves exact whole numbers, reflects the mind of a design and hence a designer. For example, in units of two Drusian feet, the station rectangle sides are 50 : 120, (the diagonal is 130 units), while in units of two AMYs, which are 5 Drusian feet in length, the Stonehenge-Preseli lunation triangle sides are 50,000 : 120,000 : 130,000. Here is 5:12:13 presented twice.

The Drusian foot is expressed here in its 'geographic' form, related to the meridian degree *(see chapter two and figure 2.8)*. The unit originates from its 'root' foot, $^{15}/_{14}$ ths of the English foot, or 1.07142857.. feet. 15:14 is a ratio indelibly stamped onto the design of Stonehenge, and remains completely verifiable. It is elegantly enshrined into the sarsen lintel circle, whose inner diameter is 97.32096 feet *(Petrie)* and whose outer diameter is 104.272457 feet *(Michell, Thom and North)*. Division of the latter by the former gives 1.07142857.., the fraction $^{15}/_{14}$ manifested at Stonehenge, and expressed as $^{30}/_{28}$ in units of Michell's royal yard of 3.4757385 feet, made up of two 'geographic' royal cubits, each a six-millionth part of the polar radius.

So, 104.272457 / 97.32096 = 1.07142857.. or $^{15}/_{14}$

97.32096 ft (inner diameter) x $^{15}/_{14}$ = 104.272457ft. (outer diameter)

97.32096 feet = 28 Royal yards; 104.272457 feet = 30 Royal yards

The Drusian foot is linked to the geographic royal cubit by the exact fraction 24/15, and thus to the royal yard by 48/15. A direct fractional relationship exists between the three key units of length identified at Stonehenge and each is fractionally related to the polar radius. Stonehenge thereby reveals its inherent metrology in a brilliant display of design which no-one has been able to appreciate until now.

Why not? Because Stonehenge is a *precision instrument* which requires that it is analysed with instruments and by people that are capable of assessing that precision. Just so long as we have assumed that the monument is a ruined pile of stones nicely arranged by barbarians, no other conclusion was possible. When

Fig 6.1 Stonehenge - a precision instrument. The photograph shows the north-eastern part of the sarsen circle, the 'midsummer' axis passing between the largest two stones of the bluestone circle, under the widest *(central)* portal and to the left of the heel stone, which is framed by the right hand portal *(photograph by Tricia Osborne)*.

Petrie and Lockyer spun their theodolites around the site, followed half a century later by Alexander Thom, a new level of precision was realised that opened the way for engineers, astronomers and metrologists to partake in a discipline that until then had been entirely organised by archaeologists.

It is now no longer tenable to maintain the view that a primitive people built Stonehenge. Seen in the perspective presented here, it is possible for the reader to endorse the enlightened view of the supremely qualified professor Thom, a man who had spent over 40 years on the ground surveying and analysing the surviving evidence from the Stone Age. In describing the megalithic builders, he said, on prime-time BBC television, that, "In terms of their thinking abilities... I think they were my superiors"*. That single opinion, a dangerous remark, shows the profound impact megalithic science can have on open minds in terms of cultural awareness.

If, after all that has been presented here, the reader should still decide that the Preseli-Stonehenge triangle is a figment of the author's imagination, then we are left with the lunation triangle as a geometric tool for predicting lunar phases and eclipses, the author thereby becoming wholly responsible for the discovery of the simplest, most exquisite geometrical device ever invented for accurately defining soli-lunar calendars and predicting eclipses.

This is assuredly not the case, for it was the Stonehenge rectangle and the revelation about the connecting geometry of Lundy and the bluestone site that first drew the author to develop the triangle for calendrical and eclipse work. And worth noting in this context is Plato's remark about the gods

Chronicle, "Cracking the Stone Age Code", BBC2,1973

choosing particularly worthless people as agents of their revelation in order to make it plain that they must be of some other origin than human contrivance.

Surely the true origin of the lunation triangle is that this simple piece of geometry describing the seemingly complex motions of the sun, moon and earth is ultimately beyond human origin. Out of all the apparent chaos and mismatch of solar and lunar cycles emerges a Pythagorean triangle which shows us mortals how to make sense of it all, even to predicting eclipses. The sacred geometry now becomes part of a cosmology, whereby sky and earth become connected.

This revelation was immured within the dimensions, the geometry and the metrology and even the choice of stones within the building we now call Stonehenge, whose siting masterfully defined a huge geometric projection of the lunation triangle across the landscape of prehistoric Britain. Through this, we have gained access into the mind of the prehistoric surveyors, and seen something of their design rules. In Book Two, John Michell reveals the extent to which this same system and these same design rules were applied throughout the length and breadth of the British Isles.

Sometime before 3000 BC, someone engineered these things. Humans just like us understood the science revealed within this book - they could even have written it. Instead of writing, however, they chose to express their wisdom concerning our planet through a symphony of numbers, ratios, fractions and geometry, a cultural flowering of sublime elegance, beauty and harmony all related to the size of the planet we share.

Understanding the system described here is the easy part. Like the measures it is rational, scientific. Understanding why this science was developed, or what effect it may have created over the population remains shrouded in mystery. As we learn to appreciate this past glory and the culture that applied it, it will demand that we look more closely at ourselves and our culture, something that is likely to be far more difficult. Our beliefs, and consequently our motives and acts, will have to adapt in order to fully embrace this newly re-discovered prehistoric science.

- End of Book One -

- BOOK TWO -
John Michell

Thanks and acknowledgements to:

Adrian Gilbert for introducing Caesar's triangle; Robin Heath for introducing the Lundy triangle and for his amiable and capable collaboration in this book; John Neal for developing the science of ancient metrology; William Stukely for perceiving works of priestly surveyors across the face of ancient Britain.

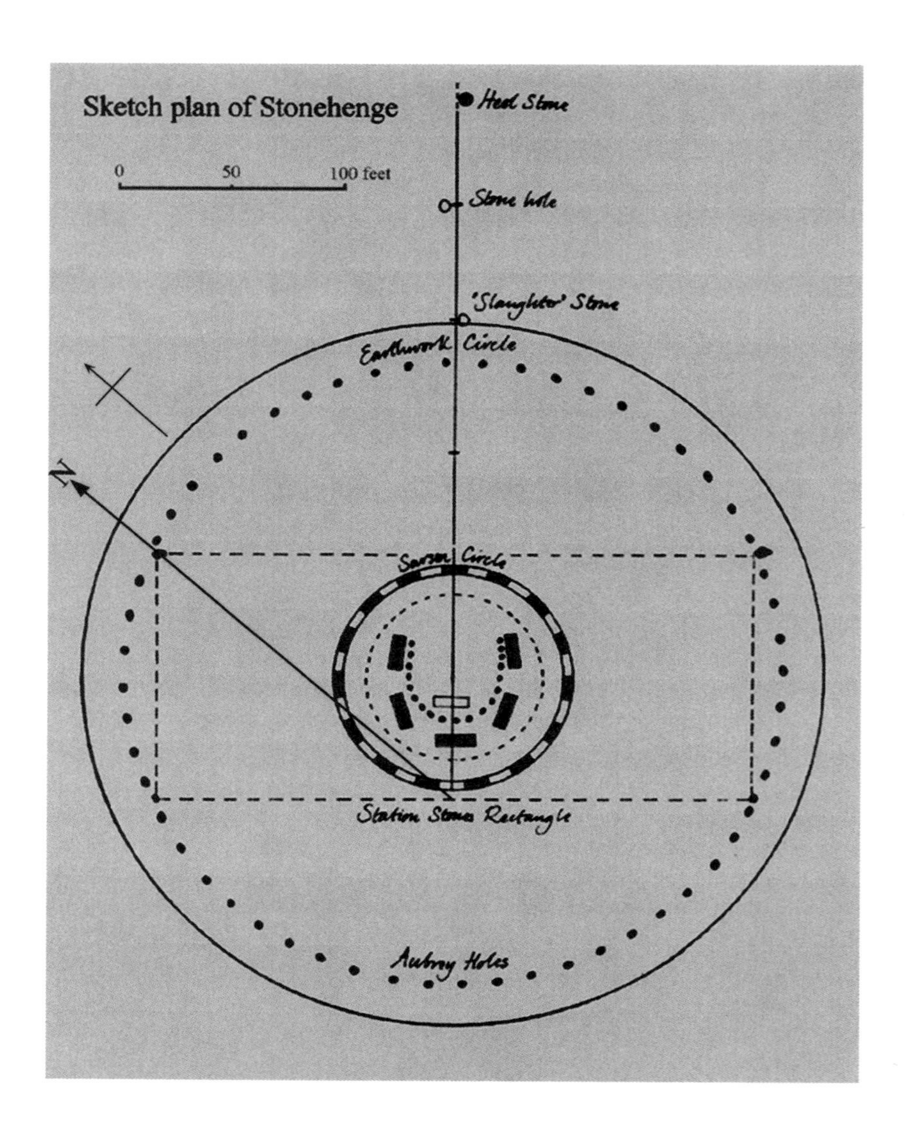

Figure 7.1 A sketch plan of Stonehenge. The building work shown here began around 3000 BC, with the construction of the circular ditch and bank, over 300 ft in diameter. The Aubrey circle, diameter 283 feet, comprised 56 large holes which may have briefly held a wooden henge structure. A few centuries later, four sarsen stones were placed around the perimeter of this circle to form a rectangular enclosure, the station stone rectangle, within which the familiar sarsen circle and five trilithons were erected, together with a bluestone circle, apparently the remains of an earlier central bluestone henge, the bluestone horseshoe and the altar stone.

Chapter Seven

Stonehenge revelations

The dimensions of the earth as known to ancient surveyors

A tradition that the whole earth was surveyed and measured in prehistoric times was acknowledged by the Greek and Roman authors and was credited up to the modern age. Sir Isaac Newton was one of the last generation to be brought up in that tradition. To prove his theory of gravitation he needed to establish the size of the earth. This was unknown at the time, but Newton understood that the ancients had surveyed the earth and had expressed its dimensions in their units of measure. This led him investigate the sacred cubit used by the architects of the Great Pyramid and the Temple at Jerusalem.

With the rise of modern science this tradition fell into disrepute and was largely forgotten. Despite the researches of Newton and the generations of metrologists who succeeded him, the lengths of the ancient units remained uncertain.

The rediscovery, beginning in 1982, of the ancient units and their exact definitions has made it possible to test the tradition of a prehistoric geographical survey - and to justify it. Not only are these units in key with the various dimensions of the earth, but they express them together in terms of a unified number code which is also a code of music and proportion.

The priestly surveyors of ancient China, Babylon, Egypt and other civilized centres used these measures to lay out vast areas of country in accordance with a cosmological pattern. Their perception was that, by reflecting on earth the order of the heavens, and through a corresponding social order, they would create a lasting abode for the gods, that is, an earthly paradise.

Land surveying developed as a sacred art, and there is evidence of it wherever people have settled. It was used in the Stone Age for locating the sites of megalithic monuments which served, among their other functions, as surveyors' landmarks. Stonehenge, the most famous of them all, stands in line with other monuments in a network of long-distance alignments set out by the ancient surveyors.

This subject became topical in the 1960s when astronomers intruded upon archaeology. They had done so before, particularly at the start of the 20th century, when Lockyer, Somerville and other scientists found relationships between stone circles and the seasonal positions of heavenly bodies on the horizon. Their observations were generally well founded, but archaeologists could not assimilate them with the evolutionist ideas of the time. Stone Age culture was held to be an oxymoron. Life then was short, brutal and primitive. Day-to-day survival was everyone's main concern. In that case, there was no time or need for scientific astronomy. Lockyer and his colleagues were called archaeologically ignorant. Their findings were disregarded, and prehistorians continued as before, untroubled by notions of ancient astronomy and science.

A basic objection to Lockyer was his picture of astronomer priests, collaborating over wide areas in surveying alignments of sites across country. How could this have been done and for what possible motives? Alfred Watkins, the founder of ley-hunting and alignment theories, encountered the same objections. So did Alexander Thom, the professor of engineering who was largely responsible for the revival of archaeoastronomy in 1967. Even when the existence of Stone Age astronomy became accepted by archaeologists, they tacitly drew the line at ancient land-surveyors.

Yet the art of surveying is wholly bound up with astronomy, particularly with the traditional form that required the setting-up of horizonal landmarks. They were two branches of the same science. Neither of them was practised in isolation; the prehistoric science to which they contributed was synthetic. It was not a secular science but religiously based and concerned with every aspect of human existence, in this world and beyond it.

That is why it is such a puzzle to the modern mind. Archaeologists admit that long barrows and other ritual monuments were to do with the cult of the dead, with oracles, necromancy and the welfare of the soul. They also recognize the orientations of these monuments, their astronomical features, their topographical relationships and the skills of transportation, engineering and masonry displayed by their builders. Each of these is a mystery in its own right, and together they constitute the great mystery of the ancient world. By our standards today, the prehistoric tribes of Britain were simple and primitive. They lived in thatched, wooden huts, along with their animals. In contrast, their tombs, temples and ritual buildings were grand and permanent, constructed with much labour, on a vast scale and to standards of craftsmanship far exceeding anything required in daily life. Everything they had and did was devoted to the gods. In return, presumably, they gained supreme benefits for themselves and their societies, the nature of which we can only imagine.

In this book are illustrated some of the major works of prehistoric surveyors in Britain. There is no theory behind this presentation. When examined, the facts speak for themselves, but they deepen the mystery rather than solve it. Yet these discoveries reveal much about ancient science and its cosmological basis. Firmly established are the principal units of measure adopted by the ancient architects and surveyors and their precise estimates of the earth's dimensions. Here also are shown the original, scientific reasons for the sanctification of certain spots, beginning with the siting of Stonehenge.

Barely touched upon, however, is the grand idea that inspired these works of surveying and geodesy. The skills and science behind them were equal to ours today, and to some extent they were evidently used for the same rational purposes, to standardize the dimensions of the earth and the lengths of its successive degrees of latitude. But there is no rational explanation for the main features of these works, their esoteric symbolism and the interplay of number, measure and musical harmonies that makes them so beautiful and delightful to study. This is more than science as we know it; it is a form of universal priestcraft, or high magic.

Stonehenge and Avebury

The two greatest, most famous ancient monuments in Britain are Stonehenge and Avebury. They are only a day's walk apart, in the same county of Wiltshire. Obviously there was a close connection between them, but they must have had separate functions because in many ways they are quite different from each other.

Stonehenge is a temple, circular and symmetrical, enclosing and protecting its sacred centre. Avebury is a sanctuary, the focus of an extensive, ritualized landscape. Ancient trackways from all over southern England converge upon it. It was the main crossroads and the centre of prehistoric life and religion. As Delphi was to the Greeks, so was Avebury to the tribes of Britain. This implies that the district around it was sacred territory or mensal land, devoted to the sustenance of priests and officials who organized the constant round of fairs,

Figure 7.2 The south-west section of Avebury showing part of the great stone circle and the dirch and bank enclosing it. The scale is shown by the figure *(right)* climbing the bank.

festivals, ceremonies and law-sessions. At the hub of it was the ritual enclosure, about 30 acres of level ground, cut off from the secular world by a deep moat and an encircling rampart of earth and chalk.

Avebury has two notable geographical features. One is its position not far from the half-way point on the main axis of southern England, the straight line between its two opposite extremities on the coast of East Anglia and at the Land's End in Cornwall. This is also the line of the old Icknield Way through the eastern counties and of the pilgrimage route between the main hilltop sanctuaries of the west country. It crosses the Avebury enclosure at its southern entrance. It is an archaic feature, recorded in legend as one of the royal roads that a long-forgotten ruler built across the length and breadth of Britain.

The second feature of Avebury's location is of more than local or national importance. Through its central enclosure runs the line of latitude that marks one seventh part of the earth's meridian, or circumference through the poles, starting at the equator. In other words, Avebury is at latitude 360/7 degrees. Its significance is therefore universal. It is the key site in Britain and the pivot of the geodetic system in which Stonehenge is a satellite.

Avebury in its beautiful landscape of green hills and watery meadows is a natural sanctuary. In contrast, Stonehenge is situated on a high plain, bleak and unsheltered. It has often been said that the site was poorly chosen. The monument would have looked more impressive on the high ground nearby, but it was placed on a lower slope and on a patch of uneven ground. This complicated the task of the builders in making the circle of lintel stones perfectly

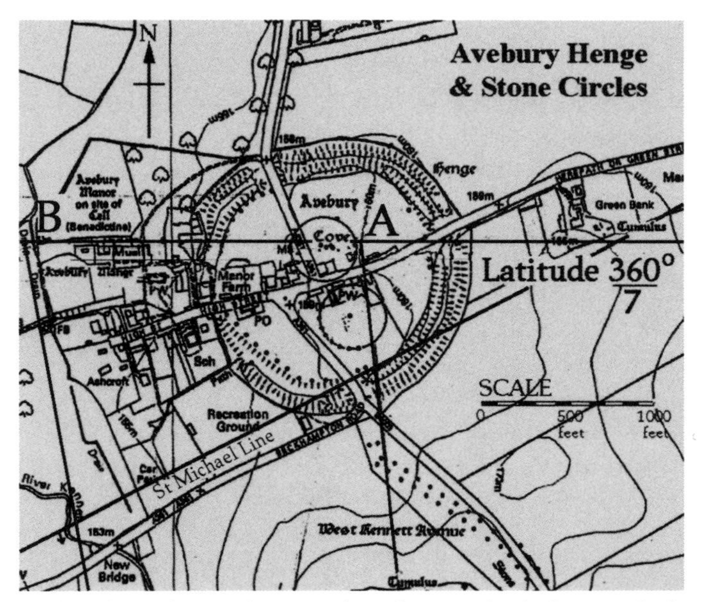

Fig 7.3 A map of Avebury, the ritual centre of southern England, showing the line of latitude 360/7 degrees that passes through it. Point A marks the terminus of the line from Stonehenge through the east end of West Kennet barrow. Point B is the terminus of the line from Stonehenge through Silbury Hill. The St Michael line or Icknield Way is also shown passing from southwest to northeast, through the southern entrance of the henge.

level. There must have been some compelling reason why that particular spot had to be the site of the temple.

The reason was geodetic - to do with surveying and measuring the earth. The position of Stonehenge relates significantly to other sites in a surveyed pattern over the island of Britain. Heath has shown its relationship to Lundy Island in the west. Also significant is its latitude. Archaeologists give this approximately as 51° 11′. Alexander Thom was more meticulous, but he was not equipped to measure fractions of a second and published the figure 51° 10′ 42″. This is near enough to the present estimated latitude of the Stonehenge centre, 51° 10′ 42.857″. It means that the distance due north to the line of latitude 360/7° is 15 minutes, or one quarter of a degree, or 17.28 miles.

It so happens that the line of latitude passing through Avebury not only divides the earth's meridian by its seventh part but also divides the 52nd degree of latitude in the proportion 3 to 4. To put it another way, Avebury stands three sevenths of the distance from the 51st to the 52nd parallel. Figure 7.5 shows the orderly way in which Stonehenge and Avebury relate to and within the 52nd degree of latitude that contains them.

On the diagram, *(figure 7.4, right)* Avebury is seen to be located 3/7ths of the distance between parallels of latitude 51 and 52 degrees, illustrating the 3:4 division. The separation between these two parallels is 69.12 miles. On the left of the diagram, this distance is shown divided into 28 parts, 7 parts of which (a quarter of the whole) comprise the distance between the Stonehenge and Avebury latitudes, with 5 parts to the south of it and 16 parts to the north. Each of these 28 divisions is 4,800 AMYs.

The length of the 52nd degree of latitude being 69.12 miles, the distance between the Stonehenge and Avebury latitudes is 17.28 miles. The section to the south, between Stonehenge and the 51st line of parallel, is 86.4/7 miles, one tenth of the distance between Stonehenge and Lundy Island.

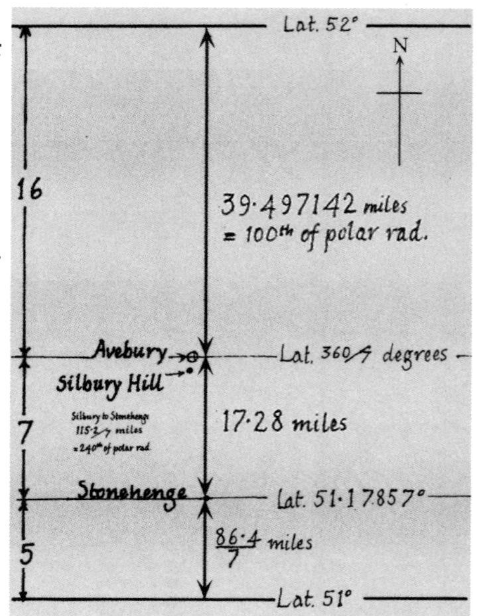

Figure 7.4 The position of Stonehenge and Avebury, in relation to the 51st and 52nd parallel of latitude. The distance from Avebury to the 52nd degree is a hundredth part of the earth's polar radius.

(For the complete diagram, see page 104).

The section to the north, between the latitude of Avebury and the 52nd parallel, is sixteen sevenths of 17.28 miles, or 39.4971428 miles - a hundredth part of the earth's polar radius *(see page 76)*.

Here are the figures.

39.497142857 miles = distance from latitude 360/7 to latitude 52 degrees;

3,949.7142857 miles = earth's polar radius.

The beautiful, simple way in which the ancient surveyors encoded this geodetic constant in their arrangement of the 52nd degree of latitude suggests that other data were included in the scheme. This is taken further in Chapter Ten.

From Stonehenge to Lundy Island

The archaic pattern of sacred geography that we are examining is so organic and 'all of a piece' that it is impossible to state for certain why any of its component sites and lines were designed. Everything, it appears, is just where it should be, as if in accordance with a grand scheme, laid down at the beginning.

One effect of the scheme is that the site of Stonehenge is precisely and ideally related to a spot at the centre of Lundy Island in the Bristol Channel. The discovery of this relationship was made by Robin Heath and first published in his *A Key to Stonehenge* (1993). He noticed that Lundy is on the same latitude as Stonehenge, and about 123.4 miles to its west. A right angle at Lundy generates a line due north that passes over the sacred island of Caldey near Tenby in South Wales and continues to the Preseli district where the Stonehenge bluestones came from. A line from there to Stonehenge completes a Pythagorean triangle with sides of 5, 12 & 13.

The only marked point in this scheme is Stonehenge. To locate the other points it is necessary to identify the unit of measure involved. This proves to be the unit of 36/7 miles which is used throughout the geodetic survey of Britain. Measured by that unit, the sides of the great triangle are 10, 24, 26, and the site of its corner at Lundy turns out to be at the very centre of the island.

Lundy is a long, narrow island, a high plateau edged with fearsome cliffs that attract the most daring rock-climbers of Britain and Europe. It is orientated due north-south and its length is precisely 3 miles. That is the length of its main axis, the straight line between its opposite extremities, north-east and south-west. The ancient geomantic practice was to locate the ritual centre of a country or island at the half-way point on its main axis. This point on Lundy is on the very same latitude as the centre of Stonehenge, and marks the spot where the line from Stonehenge terminates.

At one time Lundy Island had a native population and several villages. Traces of prehistoric settlements survive in its long-deserted northern district, and its megalithic monuments, though now scarce, were once abundant. But its ancient history is now forgotten, and its native traditions died out unrecorded. For that reason, nothing can now be learnt about its central spot, neither its name nor its former ritual function. If ancient practice was followed by the old Lundy folk, at that spot was their principal place of assembly and ritual.

Described in a previous book, *At the Centre of the World,* are the methods by which ritual centres were located, particularly in and around the British Isles. For some natural reason - geological, magnetic or whatever - many of the northern islands are elongated north-south. Examples include Shetland, the Faroes, Man, Ireland, the Hebrides and the island of Britain itself. This shape provides a north-south axis, easily identified as the overland line between opposite capes. In some cases, most obviously in mainland Shetland, the ritual centre was placed exactly at the mid-point of the axis; elsewhere it occupied a nearby site on the axis, more suitable for open-air assemblies. At that centre, kings and chiefs were installed, laws were proclaimed and justice was administered. Lundy, with its natural north-south axis, was perfectly adapted for a ritual centre that was truly central - half way along the axis at the spot on the same latitude as Stonehenge. From that spot was generated the base line of the large Pythagorean triangle.

That implies that the site of Stonehenge, in both latitude and longitude, was determined by reference to the central point of Lundy. The exact distance between the two sites is 123.42857 or 864/7 miles or 24 times the basic unit of 36/7 miles. This is where the coincidences become weird. Was Stonehenge sited in accordance with the centre of Lundy Island? or the latitude of Avebury? or one of the many other considerations, astronomical or geographical, by which theorists have explained its position?

And how is it that the natural axis of Lundy runs due north and stretches to the north-west corner of Britain, providing the main axis of the triangular figure that Julius Caesar recorded as the traditional shape of this island? As the coincidences pile up, strange thoughts arise, about the entire phenomenon of Britain and how naturally its shape and dimensions harmonize with the ideally proportioned plan that the ancient surveyors imposed on the whole island - or discovered in it.

The most interesting side of Heath's great triangle is the shortest, the side that relates as 5 to 12 with the base side between Stonehenge and Lundy. Its length, 360/7 miles, is ten times the basic unit, 36/7 miles, and when projected north to its natural terminus at Cape Wrath in the far north-west of Scotland, its length is 3600/7 or a hundred units of 36/7.

This is all so elegant that one can hardly believe it - especially since the significant degree of latitude at Avebury is 360/7 degrees.

Chapter Eight

The Numbers that Measure the Earth

Ancient geodesy is the study of the traditional dimensions of the earth and the units of measure that relate to them. These were established in very early times, as also was the practice of dividing the earth's circumference into 360 degrees.

A degree of latitude is the distance between each of the successive parallel lines of latitude, ninety in each hemisphere from the equator to the poles. These distances are different at each degree. Because the earth is not quite a perfect sphere but bulges at the equator and is flatter at the poles, the degrees of latitude increase in length the farther they are from the equator and the nearer they approach to the north or south pole.

At latitude 10 degrees, the length of the degree is 362,880 ft., easy to remember because it is 9!(called 'factorial nine') or 1 x 2 x 3 x 4 x 5 x 6 x 7 x 8 x 9. At latitude 51 degrees it has increased by the ratio 176:175 and measures 364,953.6 ft. or 69.12 miles. This is the average value of a degree of latitude. Multiplied by 360, it gives the meridian circumference of the earth as 24,883.2 miles, or a tenth part of 12^5, or 12 x 12 x 12 x 12 x 1.2 miles.

A sixtieth part of this average degree, or one minute of latitude, is 1.152 miles or 6,082.56 ft., which is the nautical mile used by navigators. The nautical unit of speed, the knot, is defined by the British Admiralty as 1.152 miles per hour.

The above figures derive from the traditional code of number, geodesy and metrology, established in very early times and used by the Stonehenge surveyors, not only in the temple itself but over long distances around it. These estimates of the lengths of degrees and the meridian circumference are as accurate as those we have today - that is, they are virtually identical. The figure of 24,883.2 miles for the circumference is better than the result during the 18th century by French scientists, who tried to measure the earth in order to establish the metre. Their measure was $28^1/_2$ miles too short.

The polar diameter is the axis on which the globe revolves. The shortest distance from the earth' centre to it's surface is the polar radius. To find its length, the meridian circumference is divided by 6.3, producing a figure for the polar radius of 3,949 $^5/_7$ miles. Twice that distance is the polar diameter of 7,899 $^3/_7$ miles. Other simple ratios link the rest of the earth's dimensions within the traditional canon of geodesy, as below:

polar radius = 3,949 $^5/_7$ miles

" " " x $^{441}/_{440}$ = mean radius = 3,958.690909 miles

" " " x $^{289}/_{288}$ = equatorial radius = 3,963.42857 miles

" " " x 6.3 = mean circumference = 24,883.2 miles

" " " x 6.3 x $^{1261}/_{1260}$ = equatorial circumf. = 24,902.94857 miles

The equator is the earth's nearest approach to a true circle. This is reflected in the value for *pi* extrapolated from these figures. It is $^{22,698}/_{7225}$ or 3.1415917, very close to its actual value, 3.14159265...

An alternative, whole-number value for the mean radius is 3960 miles. It relates to the mean radius listed above as 3025:3024, which is the ratio between two conventional values for *pi*, $^{22}/_7$ and $^{864}/_{275}$. Using the longer value for the mean radius produces simple easily remembered ratios between the three different radii of the earth, polar, mean and equatorial:

polar radius	= 3,949 $^5/_7$ miles	= $^{24}/_7$ x 1,152
mean radius	= 3,960 miles	= $^{24}/_7$ x 1,155
equatorial radius	= 3,963 $^3/_7$ miles	= $^{24}/_7$ x 1,156

The basic unit behind these measures, $^{24}/_7$ miles, is two thirds of the unit that measures the Lundy triangle, $^{36}/_7$ miles (10,000 AMYs).

The Dimensions and Metrology of Stonehenge

For many years scholars have laboured to define the lengths of the Roman, Greek, Egyptian and other ancient measures. It is an important quest, because ancient temples and precincts were designed so as to express through their dimensions the spiritual character of each site. This was done through a code of number, woven into the temple's dimensions and interpreted through the relevant units of measure.

The key to Stonehenge is in its measures. As a universal temple, it was designed after the pattern of creation, combining in its structure the numbers, ratios and harmonies by which the world was made. Also combined and made symmetrical within its plan are the movements of sun and moon. Chronological records over great periods of time are monumentalized in Stonehenge, and so are the standards of traditional metrology.

Like Solomon's temple and other national sanctuaries, Stonehenge was a repository of standards in all arts and sciences. That is why there are so many different units of measure in its design. Most prominent are the grandest of ancient units, those that constitute simple fractions of the earth's polar axis.

The principal unit of measure in Stonehenge is the width of its lintel stones. That is made plain by the design of its main feature, the thirty sarsen-stone pillars that originally supported a conjointed ring of thirty lintels. The length of each lintel is a thirtieth part of the measure round the centre of the ring (its perimeter), and the width of the lintel is a thirtieth part of the distance across it (its diameter). If the width of the lintel ring is 1 unit, the total diameter of the sarsen circle is 30 units, and its inner diameter between the inner faces of the stones is 28.

The dimensions of the sarsen circle were measured scientifically by W.M. Flinders Petrie in 1877. Despite its ruined state, he was able to establish within a small fraction of an inch its inner diameter. His figure was 97.325 ft, barely differing from the true, calculated length of 97.32096 ft. Petrie recognized this length as 100 Roman feet. He could also have called it 28 royal yards. That is the name given to the unit of 3.4757486 ft, consisting of two royal Egyptian cubits. This royal yard is the measure of the lintel width, and it represents a six-millionth part of the earth's polar radius as defined above and illustrated overleaf.

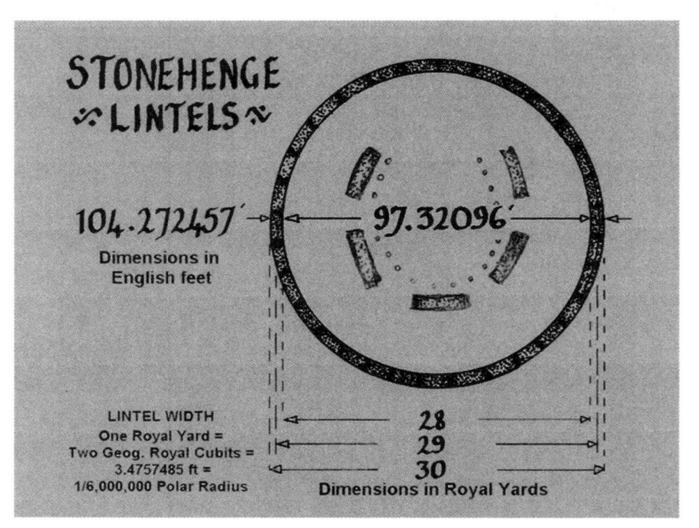

Figure 8.1 The lintel circle at Stonehenge is a level circle of curved sarsen stones, jointed together laterally by jigsaw joints and, vertically, by mortice and tenon joints involving the upright stones.

These are the dimensions of the Stonehenge sarsen circle, followed by the relationship between the royal yard and the earth's polar radius:

Stonehenge

outer diameter = 104.272457 ft. = 30 royal yards

inner diameter = 97.32096 ft. = 28 royal yards

= 100 Roman feet

one lintel width = 3.47574857 ft. = 1 royal yard

= $^{1}/_{6,000,000}$ x polar radius.

Earth

polar radius = 20,854,491.4286 ft. = 12 x 12 x 12 x 12 x $^{176,000}/_{175}$ ft.

royal yard = 3.47574857 ft. = 2 x 12 x 12 x 12 x $^{176}/_{175,000}$ ft.

As well as being measured by the royal yard, the two diameters of the lintel ring are multiples of other units. The outer diameter is both 30 royal yards and 50 sacred cubits of 2.08544914 ft. The sacred cubit, relating to the royal yard as 3 to 5, represents one ten-millionth of the polar radius.

This is the unit by which Eratosthenes, the librarian at Alexandria in the third century BC, measured the circumference of the earth. His estimate was 252,000 stades. The stade was made up of 500 feet, and the foot he was using was equal to half the sacred cubit or 1.04272456 ft. 500 of these units make the stade, and 252,000 stades give the canonical length of the meridian, 131,383,296 ft or 24,883.2 miles.

The precision with which it is now possible to define the lengths of traditional measuring units derives from the discovery in 1981 of the numerical system that structured ancient metrology world-wide. The existence of this system was hidden from previous researchers because they took the metre as their standard of reference. Invented at the end of the eighteenth century in revolutionary France, the metre was meant to represent a four-millionth part of the meridian circle - the earth's circumference through the poles. But the French survey produced a metre that turned out to be too short. Since then, the metre has several times been altered and redefined, but it still has no natural meaning and does not fit in with the system of traditional measures.

The modern metre has significant faults and disadvantages. It is not properly geodetic; it has no convenient foot or third-part division; it is awkward in use, as in making a 6-foot man measure 1.8288 m; it is not synchronized with any other known unit, now or in the past, and its definitions vary in different countries, Japan, the USA and Europe each having their version of the metre. Had the French savants who devised it adopted instead the geodetic Greek units (as was suggested at the time), we might now enjoy a scientific system of measures, universally acceptable and in tune with the true standards of ancient metrology.

The appropriate units for metrological research are the foot and the mile, now called English or (Roman) imperial. These were the basic units in the systematic code of measures whose relics are the traditional units of length in every part of the world.

The first observation that led to the rediscovery of the ancient system was of a geometric progression linking three versions of the foot, Roman, Greek and English. One of the rare data in this subject is the relationship between the Roman and the Greek foot, whereby the Roman foot plus its 24th part is equal to the Greek. In other words, 25 Roman feet equal 24 Greek feet. The Roman and Greek miles each contain 5,000 of their respective feet, so 25 Roman are equal to 24 Greek miles. The values of the classical units were closely enough known to justify the third step in the progression:

25 Roman miles = 24 Greek miles

25 Greek miles = 24 English miles.

The English mile contains 5280 English feet, so this relationship defines the other units, as listed overleaf, all expressed in English feet.

English mile = 5,280 ft

Greek mile = 5,068.8 ft.; Greek foot = 1.01376 ft.

Roman mile = 4,866.048 ft.; Roman foot = 0.9732096 ft.

Petrie was right in identifying the inner diameter of the Stonehenge circle as a measure of 100 Roman feet. The accuracy of his estimate - correct to within about a twentieth part of an inch - is almost incredible. He recognized, of course (unlike his predecessors, Inigo Jones and John Webb), that Stonehenge is far earlier than the classical Roman period, and that the various ancient units, Roman, Greek, Egyptian, Hebrew and so on, were given those names for convenience rather than implying that they belonged exclusively to any one nation. The so-called Roman foot, wrote Petrie, 'was the great Etrurian and Cyclopean unit, originally derived from Egypt, and it may have been introduced at any date into Britain.' The Stonehenge diameter is not only 100 Roman feet; it is also 96 Greek feet or 28 royal yards, and there are other units in it.

The Greek foot is the most obviously geodetic of all units. In the earth's meridian circumference of 24,883.2 miles (or a tenth part of 12^5) , there are 129,600,000 Greek feet. 1,296,000 is the number of seconds in the 360 degrees of a circle, so:

Greek foot x 100 = 101.376 ft. = 1 second of arc;

x 6,000 = 6,082.56 ft. = 1 minute of arc or 1 nautical mile;

x 360,000 = 364,953.6 ft. = 1 degree;

x 129,600,000 = $12^5/10$ miles = earth's circumference.

The Universal Language of Measure

In 1639 John Greaves, a professor of astronomy and geometry at Oxford, travelled to Egypt in order to survey, with the most accurate instruments available at the time, the Pyramids and other monuments. Having established very closely the length of the Egyptian royal cubit, he went on to Greek and Roman metrology and made similar close estimates of the classical units.

One thing he learnt was that each unit had two different values, a longer and a shorter. The longer units, those that are defined above, relate to a certain degree of latitude - the 52nd degree, where the length of one minute of arc is 6,082.56 ft (the average nautical mile) or 6,000 Greek feet of 1.01376 ft.

The shorter values of ancient metrology relate to the length of the minute at about 10 degrees of latitude, which is 6,048 ft. The corresponding degree is

362,880 ft. At that latitude the minute and degree are shorter than at latitude 52 degrees by the now familiar ratio 175:176. This gives the values of the shorter and longer versions of the Greek and Roman feet as follows:

Greek foot: longer, 1.01376 ft; shorter, 1.008 ft.

Roman foot: longer, 0.9732096 ft; shorter 0.96768 ft.

The same ratio, 175:176, produces two versions, shorter and longer, of all units in ancient metrology. Nor is that the end of the matter. J.F. Neal, the current leading authority on ancient measuring systems, claims a further range of values to each unit, up to ten in all. To some extent, at least, he is right.

Some units increase through geometric progression. For example:

1.718182 (shorter Egyptian cubit) x $^{176}/_{175}$ = 1.728 (longer cubit);

longer cubit x $^{176}/_{175}$ = 1.737874286 ft (extended cubit);

extended cubit x 2 = 3.47574857 ft (royal yard).

Independent confirmation of the Egyptian units is found in a statement by two early Greek writers, Epiphanius and Hesechius. One Roman mile, they said, is equal to 7 stades of the Egyptian foot. The Roman mile is made up of 5,000 Roman feet, and the length of that foot, as found in the diameter of Stonehenge, is 0.9732096 ft. Its corresponding mile is therefore 4,866.048 ft, the seventh part of which is 695.1497 ft. That is equal to one Egyptian stade or furlong, which consisted of 600 feet, making the length of the Egyptian foot 1.15858286 ft. The corresponding cubit to this foot is $^{3}/_{2}$ times that length or 1.7378743 ft. Two of those cubits make the royal yard of 3.47574857 ft. The Roman mile thus contains 2,800 Egyptian cubits.

In his comprehensive work on measures, *All Done by Mirrors,* John Neal lists the full range of values for each of the ancient measures, beginning with their root value and rising by regular fractions to their 'geographical' lengths, as illustrated in part one, *(figure 2.8).*

The lengths (in feet) of the Roman mile, together with the Egyptian royal cubit to which they relate, are as listed below:

	Root	Standard	Canonical	Geographical
Roman mile	4,800	4,810.909091	4,838.4	4,866.048ft
royal cubit	1.7142857	1.7181818	1.728	1.7378743ft
	(=12/7ft)	(=189/110ft)	(=12^3/1000ft)	(= half a royal yard)

All these units occur in the dimensions of ancient structures. For example, the standard side of the Great Pyramid, 756 ft in length, contains 440 royal cubits at standard value or 441 at root value. Its full height is 280 standard cubits, one tenth of a standard Roman mile. At Stonehenge, where 'geographical' values are used, the inner diameter of the lintel ring is 56 and the outer diameter 60 'geographical' royal cubits. The various Roman miles appear in the larger schemes to be examined later, including Caesar's triangle and the surveyed pattern around Stonehenge and Avebury.

Stonehenge Metrology extended

The Stonehenge lintel ring was designed in geodetic measures, notably the royal yard, and its basic relationships to the earth's dimensions are as follows:

inner radius x 2,700,000 = 131,383,296 ft = 24,883.2 miles

= earth's meridian circumference = outer radius x 2,520,000.

outer radius x 400,000 = 20,854,491 ft = 3,949 $^{5}/_{7}$ miles

= earth's polar radius = width of lintel x 6,000,000.

The unit that best expresses these geodetic dimensions is the royal yard, the sixth millionth part of the polar radius. Each of the thirty lintel stones is 1 royal yard in width. Their average lengths, however, are in terms of another unit, a very familiar one, the English mile of 5,280 ft. Divisions of the mile measure the lengths of the lintels and the spacing of the thirty upright stones on which they rest.

The average mean length of each lintel is 10.56 ft which is a 500th part of a mile. The average gap between the upright stones is a third of that, 3.52 ft, and the mean width of each upright stone at its centre is 7.04 ft, twice the size of the gap. A gap and a stone together measure 10.56 ft, the same as one lintel's length.

The mean perimeter of the ring of thirty lintels is 30 x 10.56 or 316.8 ft. which is a hundredth part of 6 miles. Dividing the perimeter by $^{22}/_{7}$ makes the mean diameter of the sarsen circle 100.8 ft. As a symbolic statement of the temple's dedication and meaning, this is the most important set of dimensions in the Stonehenge plan.

Further information on the Stonehenge metrology is provided by a contrived irregularity in the sarsen circle. The gap between the upright pillars that still frame the entrance towards midsummer sunrise is larger than the average by about a foot, and the two gaps on either side are reduced accordingly. To accommodate this feature, while retaining whole numbers in the

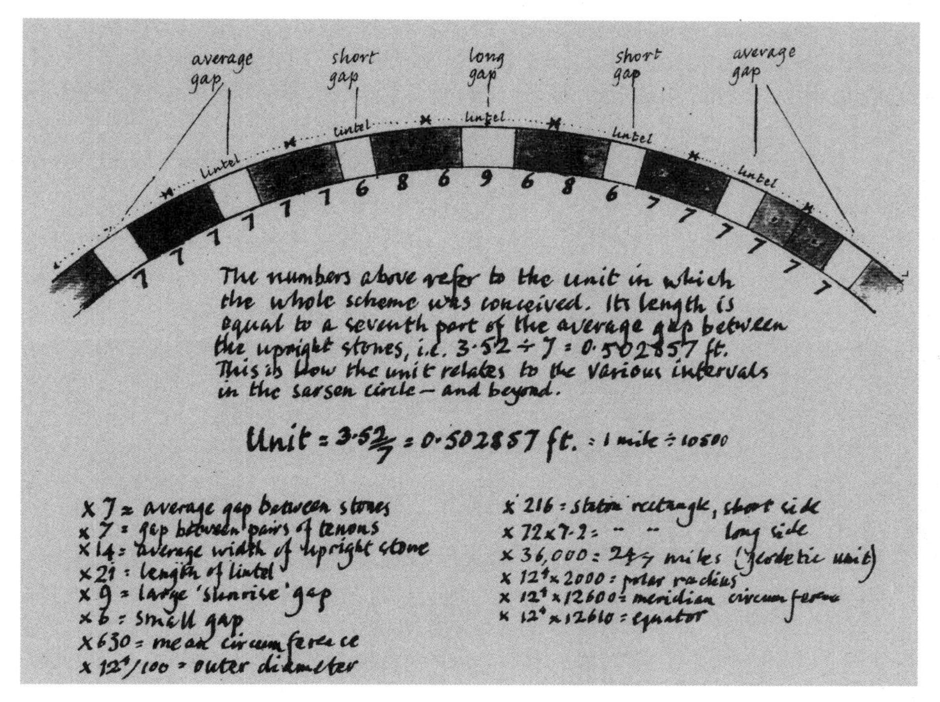

Figure 8.2 Northeast section of the lintel ring showing the larger gap between the upright stones framing the entrance. The tenons protruding from the uprights are regularly spaced at intervals of 3.52 and 7.04 feet.

dimensions, a smaller unit was required. It proves to be a seventh part of the average gap between the uprights, or one mile divided by 10,500. That makes its value 0.502857 ft. Figure 8.2 shows in plan the north-east section of the circle and how this unit applies to it.

The Stonehenge Rectangle and the Lundy Triangle.

A mysterious, geometric feature of Stonehenge is a rectangle whose corners were originally marked by short, stumpy sarsen stones. One of them lies fallen and two others have long disappeared. Their stone holes have been identified below the small mounds that now cover their sites. The corners of the rectangle are situated upon the earlier circle of 56 Aubrey holes, so that the diameter of the Aubrey circle is the same as the diagonal of the rectangle.

This figure is known as the station rectangle, its four corners being the stations. It is placed symmetrically with the lintel ring and has the same centre, so it was evidently part of the same overall plan. This rectangle plays a significant part in the Stonehenge astronomy. Its shorter sides are parallel to the circle's main axis, so they indicate the point of midsummer sunrise in one direction and of midwinter sunset in the other. The shorter sides point to the most northerly setting of the moon. Diagonal lines from the centre to the corners mark the four annual quarter days' sunrise or sunset. This discovery of a regular figure that brings symmetry into the seasons and movements of the heavenly bodies is a typical achievement of the ancient astronomers.

When the diagonals of the station rectangle are drawn on the plan, they cross in the centre at an angle of 45 degrees, more or less. This creates a geometric conundrum, illustrated below *(Figure. 8.3)*.

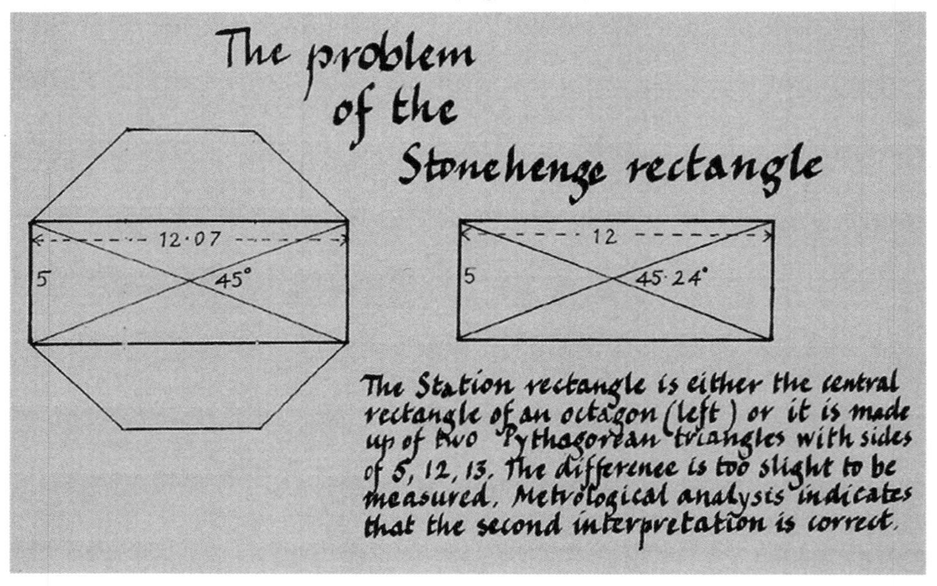

The problem of the Stonehenge rectangle
12·07
5
45°
12
5
45·24°
The Station rectangle is either the central rectangle of an octagon (left) or it is made up of two Pythagorean triangles with sides of 5, 12, 13. The difference is too slight to be measured. Metrological analysis indicates that the second interpretation is correct.

If the station rectangle was meant to represent the inside of an octagon, the angle at the centre would be 45 degrees exactly.

If it was designed as a pair of Pythagorean triangles, divided by the diagonal, with sides in the proportion 5, 12, 13, the angle would be slightly larger, 45° 14.4′.

Surveys tend to show that the angle at the centre of the station rectangle is nearer to the second figure than to the first. The most recent survey, undertaken in 1978 by Richard Atkinson, gave a figure of 45° 10′, suggesting that the 5, 12, 13 triangle was intended.

'This prehistoric rectangle', wrote Aubrey Burl, 'is near to perfection and had been laid out with great care.' According to Richard Atkinson its construction would tax the skills of any modern surveyor. It seems, however, that certain compromises had to be made in order to reconcile geometry with astronomy. The problem is to ascertain the dimensions of the station rectangle that the builders intended. To make this more difficult, three of its four original markers are no longer in place and their stone holes are large and irregular. The best modern estimates, given in feet in Thom's survey of 1973, are:

Longer sides, SW 260.0 ft; NE 260.25 ft.

Shorter sides, SE 111.1 ft; NW 110.2 ft.

The station rectangle was laid out at more or less the same time as the sarsen lintel ring. Presumably the same units of measure were used in them both. The principal unit in the lintel ring is the royal yard of ancient Egypt. Its length, as defined above, is 3.4757486 ft, and its significance is that it constitutes a twelve-millionth part of the earth's polar axis. There are 14 royal yards in the inner radius of the ring, 15 in the outer radius and 1 royal yard is the width of a lintel stone.

The estimated length of the longer side of the station rectangle, between the faces of the station stones, is 75 royal yards or 260.681143 ft. That makes the length of the shorter side, taken as 5 to 12 with the longer, equal to 108.6171428 ft. These dimensions bring out another prominent unit in the Stonehenge plan. It is a well-known unit in ancient metrology, the Drusian foot of northern Europe. Its value, confirmed by John Neal, is 1.086171428 ft. - or the earth's polar radius divided by 19,200,000. There are 120 of these units in the longer side and 50 in the shorter side of the station rectangle, a ratio of 5 to 12.

This reveals a clear-cut relationship between the station rectangle and the lintel ring. The outer diameter of the ring is 30 royal yards or 96 Drusian feet. There are 100 of those units in the shorter side of the rectangle, so there are 2 Drusian feet between the outer perimeter of the lintel ring and each of the longer sides of the rectangle that encloses it. This allows the station stones to be visible one from the other down the longer sides of the rectangle.

Another ancient unit, the Drusian 'step' of $2^1/_2$ Drusian feet or 2.7154286 ft, can be extrapolated from the dimensions of the station rectangle. 40 of those units are in its shorter side. The Drusian step is claimed by Robin Heath to be the elusive 'megalithic yard' of about 2.72 ft that Alexander Thom believed he had identified statistically in the dimensions of stone circles. The matter is controversial, but there is no doubt that the 'AMY' unit of 2.7154286 ft (= 12 x 12 x 12 x $^{11}/_{7000}$) features in the dimensions of the Stonehenge rectangle and in the great lunation triangle in which those dimensions are magnified.

The Lundy Island Triangle, its dimensions

Once the dimensions of the station rectangle are established, the way is open to investigating the great triangle with its points on Stonehenge, Lundy Island and the Preseli area of Wales.

Heath calculated that this triangle is 2,500 times larger than the 5, 12, 13 triangle formed by the diagonal division of the station rectangle. This proves to be the case, allowing the dimensions of the Lundy triangle to be defined. The figures below show how the dimensions of the longer and shorter sides of the station rectangle, multiplied by 2,500, provide the dimensions of the great triangle. These two triangles are illustrated on page 116.

longer side = 260.68114286 ft. x 2,500 = 864/7 miles

shorter side = 108.61714286 ft. x 2,500 = 360/7 miles

All these units are simple fractions of the earth's polar radius, the canonical length of which is 3,949 5/7 miles.

STATION RECTANGLE

longer side	x 80,000	= polar radius
shorter side	x 192,000	= polar radius

LUNDY TRIANGLE

Lundy - Preseli	x 76.8	= polar radius
Stonehenge - Lundy	x 32	= polar radius

If, as we suggest, the great landscape triangle determined the siting of Stonehenge, it must have been planned before the temple was founded - meaning that it is at least 5000 years old. The Stonehenge architects miniaturized its dimensions in the station rectangle.

Chapter Nine

Traditions of Ancient Surveyors in Britain

The surface of Britain is like a palimpsest - an old parchment covered with lines and markings from different periods over thousands of years. Its most obvious feature is the modern road system, but below it, and almost invisibly extending it, are straight lines of ancient tracks or boundaries, linking prominent landmarks and running between opposite ends of the island.

These mysterious trackways are explained in early British chronicles as the works of great rulers in the past. One of these was Elen or Helena in the 4th century AD, a Celtic Christian who married a Roman emperor, grandly named Magnus Maximus, and appointed her many sons to rule different regions of Britain. Her road-building achievement is described in *The Dream of Macsen Wledig*, an old Welsh story included in the Mabinogion.

'Elen decided to construct highways from one castle to another across the island of Britain. And the roads were made. And for that reason they are called the roads of Elen of the Hosts, because she was sprung from the island of Britain, and the men of Britain would not have come in hosts to do the work for anyone but her.'

Stretches of ancient stone roads in Wales are known as Sarn Elen, Helena's causeway. They predate their supposed builder, so the Elen they were dedicated to is likely to be the old Celtic spirit, corresponding to Hermes, who was the guardian of tracks and travellers.

Another famous road-maker was King Belinus, ruler of all Britain in the fifth century BC. The story is told by Geoffrey of Monmouth in his *History of the British Kings* (Lewis Thorpe's translation),

'He summoned workmen from all over the island and ordered them to construct a road of stones and mortar which should bisect the island longitudinally from the Cornish sea to the coast of Caithness and should lead in a straight line to each of the cities on the route. He then ordered a second road to be built, running west to east across the kingdom from the town of St Davids on the Demetian sea over to Southampton and again leading directly to the cities in between. He built two more roads in a diagonal pattern across the island, to lead to the cities for which no provision had been made. Then he consecrated these highways in all honour and dignity, proclaiming it to be an integral part of his code of laws that punishment should be meted out to any person who committed an act of violence upon them.'

Geoffrey continues: 'If anyone wishes to know the full details of the highway code established by Belinus, he must read the Molmutine Laws which the historian Gildas translated from Welsh into Latin, and which King Alfred later re-wrote in the English language.'

It is made plain in the text that Belinus did not so much make new roads as repair and restore order to an old highway system that was falling into ruin. The reason for his work was that the old tracks had wandered off line, and 'no one knew just where their boundaries should be'. Nor was he the first to sanctify the old roads and make highway robbery a crime of sacrilege. His father, Molmutius, author of the famous Laws, codified the traditional right of travellers on the sacred paths between the national sanctuaries. Every wayfarer - criminal, debtor or fugitive - was free from harassment or arrest. According to the enactment, a child or a foreigner with no knowledge of our language could pass from one end of the country to the other, finding hospitality everywhere.

Reflected by these laws is a very beautiful kind of society - peaceful and highly cultured. They awaken innate memories of a lost time and country, not so very long ago, when the Holy Grail was apparent on earth and life was experienced on a glorious, mythological level. That state is called an enchantment. It is an induced state, the creation of an initiated priesthood. The art by which it is implemented is priestcraft, and its most powerful agent is music. Hearing the same classical modes of music at the same places at regular festivals throughout the year, the country people were held under a religious spell. Successive generations lived the same way as their ancestors, maintaining the same high standards of culture and husbandry. In that timeless existence the minds of thinkers were drawn upwards towards the ideal, and their energies were engaged in binding heaven ever more closely to earth. One expression of that is the pattern of surveyors' marks, tracks and alignments that rationalizes the topography of Britain and displays the whole island as a harmonious creation. In that pattern lies the secret of the Grail, and in its modern revelation is the sign that prophecies of the Great Return will be fulfilled - soon or in God's good time.

The Old Roads of Britain

The subject of ancient rulers and their long-distance roads across Britain is puzzling and controversial. There are such roads, and long stretches of them were incorporated in the Roman system, but other track-lines were unrelated to Roman activities. It is a confusing pattern. Some old roads run straight and strong to a certain point, where they change course or end abruptly; some were laid out in parallel lines or at precise right angles, and some point to alignments of churches and landmarks but do not continue through them. Most interesting are the unconnected lengths of roads, paths and local boundaries that, with many gaps and deviations, conform to the same line from one end of the country to the other.

An obviously surveyed road-pattern is centred upon High Cross on the borderline between the counties of Leicester and Warwick and at the junction of four parishes. A monument there marked the crossing-point of two long-distance roads, Watling Street and the Fosse Way. These are possibly the roads that, in Geoffrey of Monmouth's account, formed 'a diagonal pattern across the island'. The course of Watling Street, from the Kent or Sussex coast to the north-west, is made up of straight lines and angles. The Fosse Way is more direct, barely deviating from a straight line. Starting on the south-west coast at the point where Devon meets Dorset, the line proceeds to High Cross by way of Ilchester, Bath and other Roman centres. Continuing north-east by way of Lincoln, the Fosse Way has its terminus at the estuary of the Humber.

The location of High Cross and its proclamation as the centre of Britain was evidently Roman work. Its site was established by the methods of augury, the mystical code of astronomy and land surveying practised throughout the Empire. The key feature of this art was the 'Cardo', the line running north-south between the opposite ends of an island or of any integral territory. This represented the fixed pole or axis from which all measures were made.

Figure 9.1 *(above)*. William Stukely's engraving of 1723, shows the ancient trackways that he knew from his antiquarian journeys through Britain. Prominent is the line of Icknield Street through the eastern counties. Extended westward to Land's End it forms the main axis of southern Britain, centred upon Avebury. The dashed line shows its route as accurately as this map allows.

Figures 9.3 & 9.4 *(opposite)*. The meeting place of Watling Street and the Fosse Way, drawn by William Stukeley in 1728, was the site of the Roman High Cross.

The farthest point of Roman dominion in Britain was at one time the eastern, coastal terminus of Hadrian's wall. There, by South Shields, a fort was erected, and from that point a line was surveyed southward to the opposite extremity, St Catherine's Point at the tip of the Isle of Wight. The lines of Watling Street and the Fosse Way were adapted so as to meet upon the Cardo at angles of 30 degrees each side of it. At that spot the Roman surveyors set up the High Cross and around it they laid out a town, Venonae. Also placed on the axis was the great Roman military centre, Templeborough in Yorkshire.

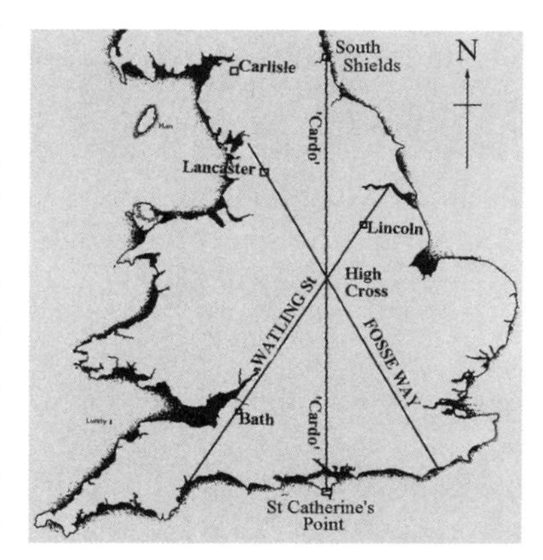

Fig 9.2. The symmetry of Watling Street and the Fosse Way, whose paths were adapted to meet at High Cross.

Reasons why that particular site was chosen to be the centre of Roman Britain, rather than any other spot around the mid-point of the axis, are apparent at the site itself. It is a natural beacon site and widely visible. It was said that 57 churches could be seen from the High Cross.

The Britain of which the High Cross marked the centre was not the island as a whole, but its southern, Roman section below Hadrian's wall. The whole scheme must date from 122 AD when the wall was begun. The purpose of that structure was primarily geomantic. According to the Encyclopaedia Britannica, 'It was to be, as it were, a Chinese wall, marking the definite limits of the Roman world'.

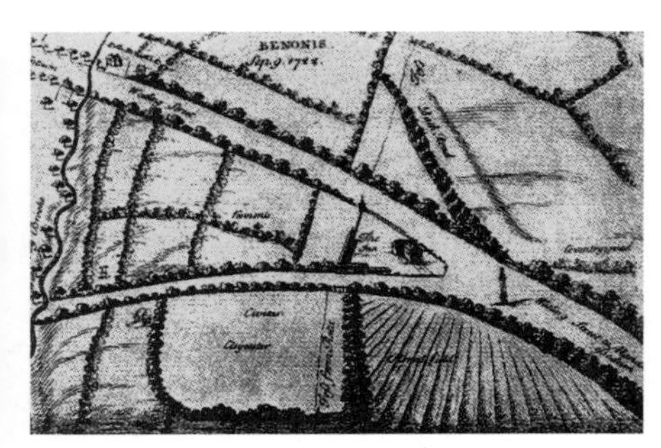

Fig 9.3

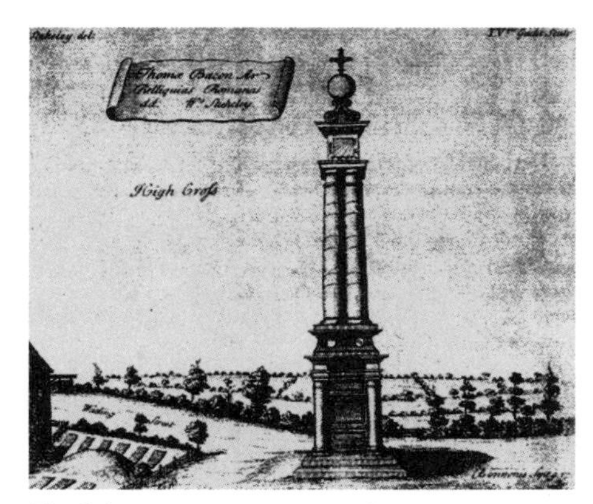

Fig 9.4

The four 'royal roads' to which the rights of sanctuary extended were specified in the Laws of Edward the Confessor as Watling Street, the Fosse, Ermine Street and Icknield Street. The first two make the X pattern over England; Ermine Street is the old north-south route through London towards York, and the line of Icknield Street runs across southern England, between its opposite extremities, the Land's End of Cornwall and the coast of East Anglia. It is popularly known as the St Michael line.

None of these roads can be identified throughout its entire length. There are spurs, deviations and long stretches that have been lost or were never built. The Romans, despite their reputation for taking the shortest path between points a and b, never did so over long distances. They adapted the existing British roads as long as they led in the right general direction. Modern road-builders have largely followed their pattern. With these recent overlays, the original pattern of the ancient British roads is almost obliterated.

The key to its rediscovery lies in the original description of the four royal roads. They were laid out as straight lines from one end of the island to the other, up, down and across. This obviously implies a national survey. Belinus's first undertaking was clearly the work of a surveyor. He drew a straight line to "bisect the island longitudinally from the Cornish sea to the coast of Caithness". In other words, he drew the famous line between John o' Groats and the Land's End, the main axis between the two opposite extremities of the British Isles, with its centre at the former Archdruid's establishment on the Isle of Man.

The picture that is emerging is of an ancient survey of Britain which not only linked the most distant points of the island but created a large, regular geometric pattern across it. The surveyors established their lines and then, in relation to these, sanctuaries were located and communities developed around them. Sections of the surveyors' lines, those that ran between the settlements, were marked out as tracks and paved roads and were declared sacred. Elsewhere the lines were represented by alignments of shrines and monuments. This can be seen most clearly on the course of Icknield Street, one of the major lines that were laid out between opposite ends of the country.

Icknield Street and the St Michael axis

Here is the definition of Icknield Street, from Bartholomew's *Survey Gazetteer of the British Isles*, 1904:

> "Icknield Street, ancient Roman road, crossing from E. to SW. of England; began in Norfolk and terminated at Land's End."

This brief description was drawn from researches by antiquarians of the 17th century and later. One of them, Dr Robert Plot, author in 1677 of *The Natural History of Oxfordshire,* made a special study of the four royal roads, concentrating on the Icknield Way that ran through his county. He traced its line from East Anglia to the Land's End, and derived its name from the Iceni tribe of eastern England who famously resisted the Romans under Boadicea.

It is now known that the Icknield Street or Way is long pre-Roman, and only short sections of it were adapted to Roman use. The original line, straight from the coastal border of Norfolk and Suffolk to Land's End, can still be drawn, exactly as the ancient surveyor ruled it. It has been much explored and written about since its rediscovery in 1969, and has been called the St Michael line - or axis.

This line is very archaic. It begins in nature, as a fact of geography, being the longest line that can be drawn through southern England and thus providing its axis. Nature has also ordained that certain prominent landmarks, the dramatic cone of Glastonbury Tor for example, should fall precisely on the line. These coincidences evidently impressed the ancient surveyors, for they sanctified this line as probably the first of the royal roads through Britain. Near its mid-point at Avebury they established the national sanctuary, the British Delphi. Two reasons for that choice have already been noted: that

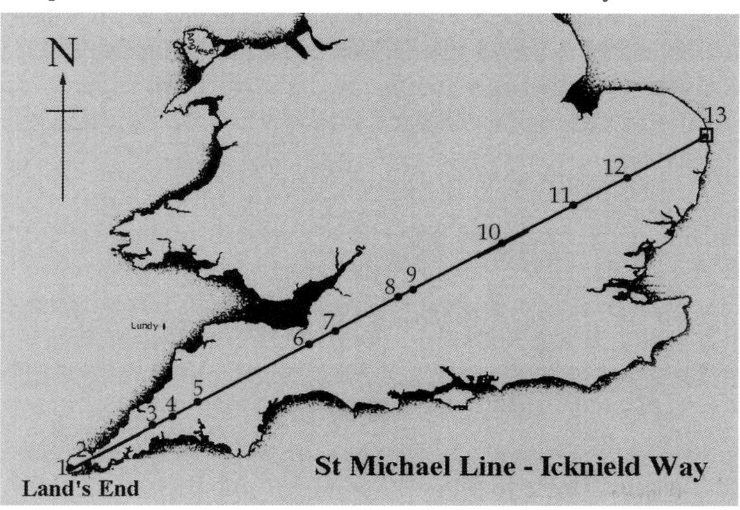

Figure 9.5 Icknield Way, often referred to as the St Michael line. Key: 1. St Michael's, Carn Brea. 2. St Michael's Mount. 3. St Michael's Chapel, Roche Rock. 4. The Cheesewring. 5. St Michael's, Brentor. 6. St Michael's, Burrow Mump. 7. St Michael's, Glastonbury Tor. 8. Avebury. 9.St George's, Ogbourne St George. 10. Ivinghoe Beacon and existing stretches of Ichnield Street. 11. Royston Cave. 12. Abbey, Bury St Edmunds. 13. Norfolk-Suffolk Border, near Hopton.

Avebury stands upon the line of Icknield Street and at the significant latitude, 360/7 degrees. Another reason was the benevolence of nature in shaping Avebury and its surroundings as an imitation of paradise. Seven rivers have their sources there, and there is a quality in its light and atmosphere that is apparent at other sacred places. With its chalk downlands and fertile meadows, Avebury today is still one of the richest areas of Britain.

East and west of Avebury the line of Icknield Street has two very different characters. To the east, the straight line contines through the Chilterns and Ivinghoe Beacon, to Bury St Edmunds and along the Norfolk-Suffolk border to the meeting point of those counties on the North Sea. That is precisely the main line of the Icknield Way. As a road, the Way is ill defined. It wavers each side of the line and sometimes divides into two pathways on either side of it. East of Avebury it deviates southwards to a convenient crossing of the Thames at Goring, approaching the central sanctuary by another ancient track, the Ridgeway. In East Anglia the Icknield Way has many spurs, leading to places of local importance. Yet overall, despite its spreadings and wanderings - largely for the convenience of the cattle-drovers who used it - Icknield Street adheres to the straight line that King Belinus drew along the axis of southern Britain.

The least defined part of Icknield Street is its eastern end, between Bury St Edmunds and the sea. The road branches off in different directions and the main line is no longer apparent. But it has left its mark in the names of old farms and villages along the route. Among them, moving from east to west, are: Ditchingham, Dickleborough, Rickinghall, Ickworth, Ickleton, Ickleford, echoing the name of the line on which they were founded.

Did the Icknield Way take its name from the Iceni, or vica versa, or was there no connection? That question has caused much argument among antiquarian scholars. Some have pointed to ancient tracks called Icknield, Ryknield or Hackney that occur in other parts of the country, suggesting that the Ick name is generic and meant a certain kind of sacred route. The Via Egnatio, running from Rome to the eastern capital of the Empire at Constantinople, has a similar sound. A learned guess is that the name derives from the Greek, *ichnos* meaning a trackway or the footprint of a traveller.

To the west of Avebury, where it passes through the southern entrance, the line is continued for some miles by the old road to and beyond the Beckhampton roundabout. Between there and Glastonbury Tor no track is apparent, but the line runs along the axis of the Tor and the old pilgrims' path that crosses it.

About ten miles further west the line goes down the axis of another remarkable hill, Burrow Mump, at Burrowbridge. Perched on its summit is a church,

Figure 9.6 Barrow Mump. lying on the St Michael alignment ten miles from Glastonbury Tor.

dedicated to St Michael, now ruinous. As its name implies, the Mump is a barrow - a large, piled-up heap of earth. Usually that means an artificial mound, but the Mump is huge - far bigger than Silbury Hill or any other known earthwork. For that reason it is supposed to be a natural feature. If so, its occurrence on the same axis line as Glastonbury Tor is an amazing coincidence. It is the right hill, in the right place and at the right angle. All around it the country is flat marshland; it is the only hill within miles. It stands like a fortress beside a river, where the axis line and the modern road both cross it. Standing on Burrow Mump (*above*) and looking east to Glastonbury Tor on the skyline, its is hard to accept that this sighting-point of the direct line to Avebury happens to be there by chance. It looks in every way like a gigantic work of the ancients.

The same mystery recurs further west, on Bodmin Moor, where the axis line passes over another enigmatic structure, the Cheesewring (*shown overleaf*). It is a tall pile of huge, flat rocks, balanced on a hilltop. The old Cornish antiquarians, influenced by local legends, identified it as an altar erected by the Druids. There is good evidence that it featured in ancient rituals. Modern geologists consider the Cheesewring to be a natural formation, though not easily explained. This question is opened again by its position as a marker upon the Icknield axis.

In its course through the West Country, the Icknield line has a different character from its eastern section. The Ick names no longer occur and there is no obvious long-distance trackway. The line seems like a mystical pilgrimage route. The sites that it passes through or by are the prominent rocks and hilltops that in Celtic lands are dedicated to the Archangel Michael. On these high sanctuaries hermits kept a light for the guidance of travellers, and churches or pilgrim chapels were later built upon them. Directly on the Icknield line are St Michael's church on the Tor, St Michael's on Burrow Mump and The

Figure 9.7 The Cheesewring, directly on the St Michael line and poised on a high outcrop on Bodmin Moor.

Cheesewring in the west of Cornwall; while beside it stand the church of St Michael on Brentor, a steepling rock on the edge of Dartmoor; St Michael's rock, chapel and hermitage at Roche, and the impressive island rock of St Michael's Mount in the far west.

This western half of the Icknield line has nothing to do with the Romans but retains the spiritual quality of the ancient British survey. The cult of angels and archangels was deep-rooted in the Celtic religion, and it remained so after the reformation that replaced Druidry with the Celtic form of Christianity. From Britain it was brought by missionaries into the continent and the Church of Rome. It was a mystical process, and in some way the sacred line to Glastonbury from the Land's End and St Michael's Mount had a part in it. In West Country legends and folklore it was the route that Jesus took from Cornwall to Avalon, or Glastonbury, and its college of learned Druids.

In previous books, beginning in 1969 with *The View over Atlantis* and more accurately in *Twelve-Tribe Nation*, this axis line was called the St Michael line. Much has since been written about it under that name. It was explored on the ground by Paul Broadhurst and the dowser, Hamish Miller, and described in its mystical, dynamic aspect in their book, *The Sun and the Serpent*. The same writers then produced an even greater book, *The Dance of the Dragon*, about their pursuit of another line of major St Michael sanctuaries, from St Michael's rock off the coast of Ireland, through St Michael's Mount, Mont St Michel and a string of archangelic sanctuaries to Greece and, finally, Mount Carmel in the Holy Land. If this implies an ancient surveyed line, as seems likely, the system that occurs in Britain extended far wider than this island - was indeed universal.

Bartholomew and earlier authorities defined the St Michael line, as it is now known, but they called it Icknield Street and identified it as one of the sacred roads of a legendary ruler. Since their name is old and traditional it is the proper one. The St Michael axis was a useful name in its time, but it is correctly and more significantly referred to as the Icknield line. It is the most interesting and easily studied of the original lines in the great survey, or mystical enchantment, that the initiated rulers of ancient times laid across the island of Britain.

CAESAR'S TRIANGLE

Confirmation that the whole of Britain was surveyed in early times appears in Julius Caesar's *Commentarii de bello Gallico* of 51 BC. This is what he says about the shape and dimensions of the island:

'It is triangular, with one side facing Gaul. One corner of this side, on the coast of Kent, is the landing-place for nearly all the ships from Gaul, and points east; the lower corner points south. The length of this side is about 475 miles. Another side faces west, towards Spain. In this direction is Ireland, which is supposed to be half the size of Britain, and lies at the same distance from it as Gaul. This side of Britain, according to the natives' estimate, is 665 miles long. The third side faces north, no land lies opposite it, but its eastern corner points roughly in the direction of Germany. Its length is estimated at 760 miles. Thus the whole island is 1,900 miles in circumference.'

The lengths of the three sides are 475, 665 and 760 Roman miles. Dividing each of them by 95 reduces the dimensions of Britain to a simple triangle with sides of 5, 7 and 8. Adrian Gilbert studied this figure and noticed some interesting features. From the apex A *(figure 9.8, overleaf)*, a line drawn perpendicular to the base line at D divides BC into two sections so that BD = 1 and DC = 4. Another perpendicular, from B to E, divides side AC into $5^1/_2$ and $2^1/_2$ units. The angle at A is 30 degrees, so it follows that the angle EBC must also be 30 degrees.

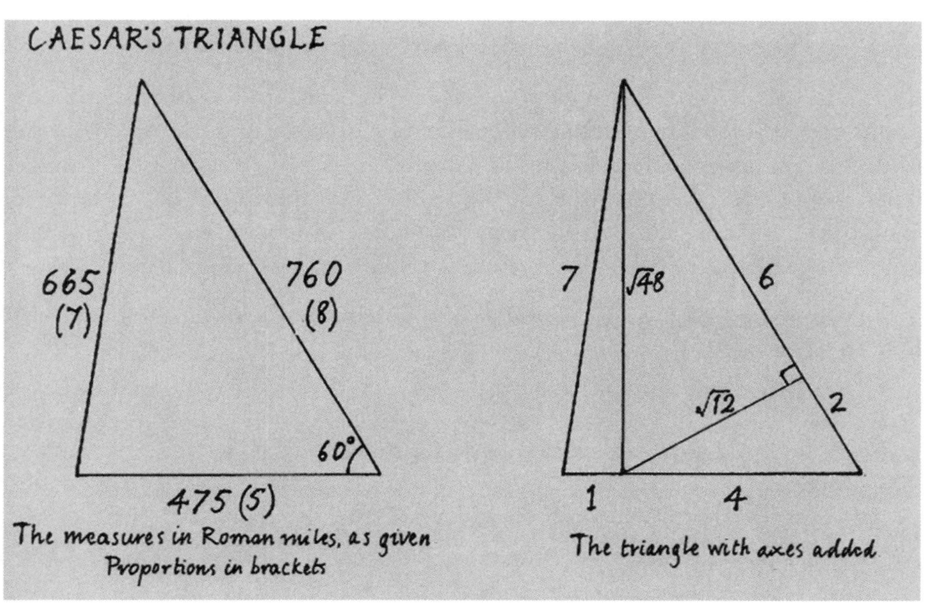

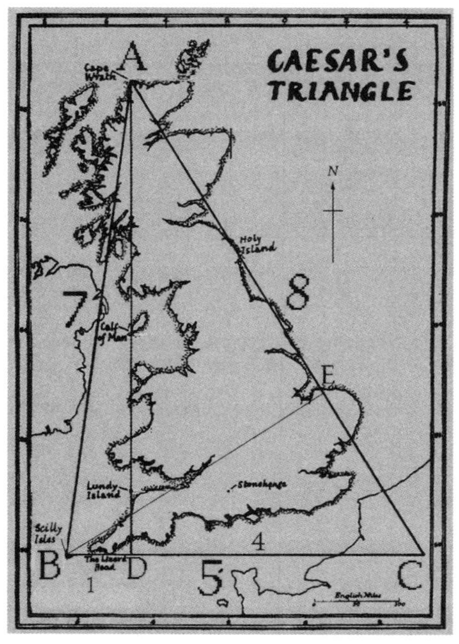

Fig 9.8 Evidence of a pre-Roman survey of Britain, according to Julius Caesar in 51 BC. The dimensions are units of 95 Roman miles, or 86.4 English miles or 168,000 AMY.

The feature in this triangle that allows it to be placed accurately over the map of mainland Britain is its perpendicular, AD. It is clearly identified as the north-south axis, the line from Cape Wrath, the most northerly point of north-west Scotland, to Britain's southernmost line of latitude, at the tip of the Lizard in Cornwall. It passes through two key sites in the ancient survey of Britain, the centres of Lundy Island and the Calf of Man. These islands are exactly 200 miles apart, a third of the whole axis. At Lundy this axis makes a right angle with the Lundy-Stonehenge baseline of Heath's lunation triangle, and from Lundy to Cape Wrath the distance is ten times the shortest side of that triangle, or $3600/7$ miles.

This axis runs due north-south, not following the earth's curved surface like a meridian line, but parallel to its polar axis. It seems to have been drawn straight across a map of Britain, or laid out across the land as if the island lay upon a flat plate.

The length of the axis is evidently intended to be 600 miles. That is an appropriate measure for the symbolic pole, 6 being a number of authority, as in the 6-foot rod or sceptre held by a traditional ruler. It extends over 8° 40′ 15″ of latitude, from Cape Wrath at 58° 37'30″ to Britain's southernmost line of latitude, parallel to the Lizard Head. The southern tip of Lizard Point is 49° 57' 15″ making the actual length 598.64 miles, which is 440 : 441 of 600 miles.

Taking the length of the vertical axis as 600 miles, the other dimensions of Britain's triangle can be calculated as below.

South side, 432 miles, divided into two sections of 86.4 and 345.6 miles;

West side, 604.8 miles; East side, 691.2 miles;

The measure round the three sides of this triangle is 1728 or 12 x 12 x 12 miles, and the common unit in its dimensions is 86.4 miles.

These figures are not the same as Caesar's but differ from them by the ratio 10:11. That means that the mile he used was the shortest, root version of the Roman mile, 4800 ft, relating to the 5280ft English mile as 10 to 11. This and other values of the Roman mile are listed on page 81. Measured by the root Roman mile, the dimensions of Britain, as represented by this triangle are as listed below:

South side, 475.2 Roman miles, divided as 95.04 and 380.16 Roman miles;

West side, 665.28 Roman miles;

East side, 760.32 Roman miles;

Perimeter, 1,900.8 Roman miles.

The south side, according to Caesar, is 'about 475 miles'. He omitted the decimal or fractional additions that complete the figures. In units of the 4,800-foot Roman mile, Caesar's triangle has a perimeter of 1,900.8 or 12 x 12 x 12 x 1.1. In units of the English mile, the perimeter becomes 1,728 or 12 x 12 x 12, and the three sides are 432, 604.8 and 691.2. In terms of the AMY (19.008/7 feet) identified by Robin Heath as the megalithic yard, these lengths are respectively: south side, 840,000 AMY, west side 1,176,000 AMY, east side 1,344,000 AMY, perimeter 3,360,000 AMY. The perimeter is exactly 25 meridian degrees, each of 69.12 miles.

This is not just a solution to the old problem of Caesar's triangle, but the key to a far more ancient secret - the forgotten science of prehistoric surveying. This triangle forms the outer framework of many other, more detailed and local surveys. The overall picture is of an entire country overlaid with a series of geometric patterns, linking its geographical extremities with landmarks and monuments and, at the same time, expressing astronomical data and the intervals of music.

These discoveries can be called revelations. They are wonderful in themselves, but what they indicate is something even more wonderful - the existence of a further mystery behind these relics of prehistoric culture, the mystery that has been called the Matter of Britain. Behind these preliminary studies are further dimensions of knowledge and wisdom, to which we have no access. This is not meant to discourage future researchers - quite the opposite. This is the most delightful and rewarding of subjects. The more closely you inspect the works of the prehistoric surveyors, and enter thereby into their minds, the nearer you approach the mysteries of ancient Britain in its long-lost state of magical enchantment.

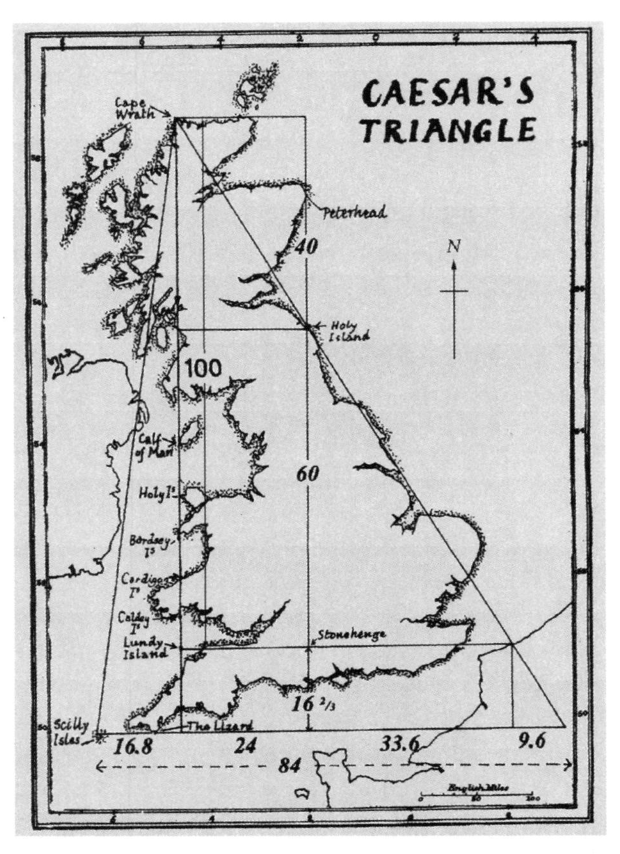

Figure 9.9 Caesar's triangle overlaid onto the 'ladder' which formed part of the prehistoric survey of Britain. The unit of measure marked on this map is $^{36}/_{7}$ miles, the measure fundamental to the whole survey. The 'rungs' of the ladder are 24 such units in width, which is $^{864}/_{7}$ or 123.42857 miles, the distance from Stonehenge centre to Lundy centre.

The Harmonic Divisions of Britain

The pivot of Caesar's triangle is its north-south axis through Lundy Island. Intervals along it are marked by certain spots and islets in accordance with a numerical code which can be read as geometric proportions or a scale of music.

This axis runs from Cape Wrath at the northernmost point of Caesar's triangle to the latitude of the southernmost cape of Britain. According to nautical tradition, as illustrated in William Borlase's *Antiquities of Cornwall* (1st ed., 1754), Britain's southernmost line of latitude runs through the largest of the Scilly Isles, St Mary's, and a point just off the Lizard Head in Cornwall. From that latitude (here defined as 49° 56′ 18.57″) to Cape Wrath the distance

Figure 9.10 The Calf of Man. *(photograph courtesy of Manx National Heritage)*

is 600 miles. The most outstanding divisions along its course are marked by the centres of Lundy Island and the Calf of Man. These two small islands are 200 miles apart, so the three intervals are as listed below.

Cape Wrath to Calf of Man = 314.2857143 = $2200/7$ miles;

Calf of Man to Lundy = 200 miles;

Lundy to southern end of axis = 85.7142857 = $600/7$ miles.

The common unit behind these three measures is the 21st part of the axis, $600/21$ or $28\,4/7$ miles. Applying this unit to the distances above,

Cape Wrath - Calf of Man = 11 units;

Calf of Man -Lundy = 7 units;

Lundy - southern end = 3 units.

These divisions imply a harmonic scale. They also provide the principal ratios in this great survey, beginning with the distance from Cape Wrath to the Calf of Man, $2200/7$ miles. This is easily identified as the very first ratio in geometry, being a hundred times $22/7$, the usual approximation to pi. Add to that the Calf-Lundy distance of 200 miles, and the interval between Cape Wrath and Lundy is seen to be $3600/7$ miles. This is a hundred times the unit of $36/7$ miles that features throughout the ancient British survey. The third division, southwards from Lundy, is of 600/7 miles, or a seventh part of the whole axis.

This is merely the outline of a far more elaborate structure whose basic feature is a ladder or scale (échelle) up the length of Britain *(see appendix five)*. Its left leg is the vertical axis through Lundy; its right leg is the parallel line through Stonehenge and up through Holy Island off the coast of Northumberland. The rungs on this ladder, measuring 864/7 miles between the two legs, mark the intervals in the scale.

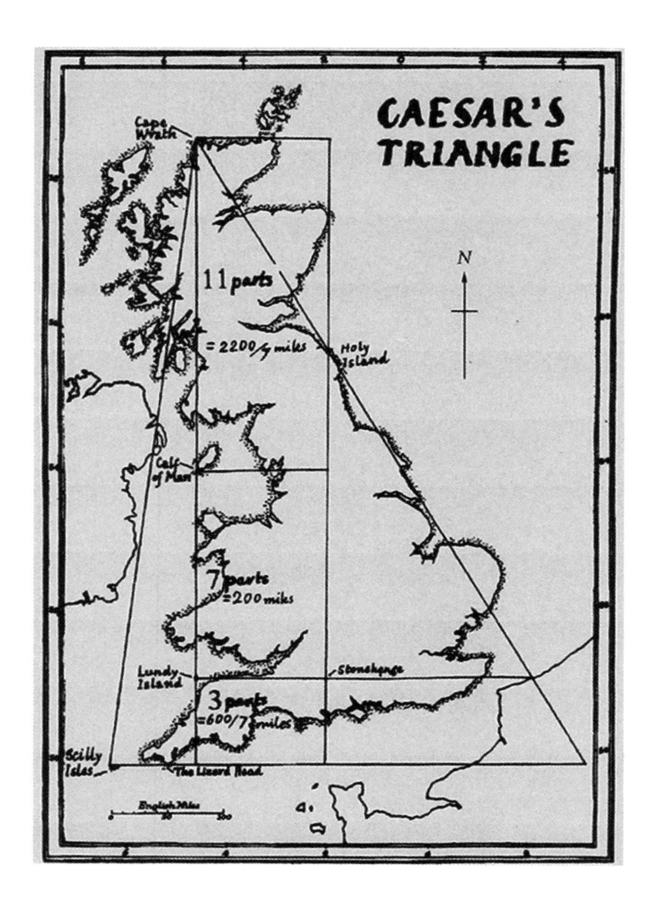

Figure 9.11 The harmonic 3, 7, 11 spacing of the axis of Caesar's triangle. The axis length, from Lizard Head to Cape Wrath, totals 600 miles. The distance fom Lundy to the Calf of Man is 200 miles, while the distance from Lundy to Cape Wrath is 3600/7 miles.

Chapter Ten

Secrets of the 52nd Parallel

This chapter resumes the subject of the geodetic and other significant lines in the Stonehenge-Avebury region and within the 52nd degree of latitude. There are three such principal lines. They begin at Stonehenge and proceed northwards, to the latitude of Avebury and then further to the line of latitude 52 degrees. In relation to these lines are structured many of the large-scale surveyors' patterns that occur over this district. The three lines are:

(i) The meridian, the direct line projected due north from Stonehenge;

(ii) The line from Stonehenge, through the east end of the West Kennet long barrow to Avebury;

(iii) The line from Stonehenge, through Silbury Hill.

These three lines yield through their measurements the lengths of the three radii, polar, mean and equatorial, of the terrestrial globe - unequivocally, in clear numbers and in agreement with the traditional code of geodesy.

The first line, due north, runs parallel to the main axis through Lundy Island of Caesar's triangle and is directed to Holy Island off the coast of Northumberland. The section, running north from Stonehenge to latitude 360/7 degrees on the ancient Ridgeway track east of Avebury, is 17.28 miles long, or a quarter of a degree of latitude. From there to latitude 52 degrees the distance is 39.4971428 miles, a hundredth part of the polar radius.

The second line runs at an angle of about 4 degrees west of the first, from Stonehenge to the significant line of latitude at Avebury, passing through the great stones at the east end of the West Kennet long barrow. Its calculated length is 17.3192727 miles, exceeding that of the first line by 207.36 ft. In the fifteen minutes of degree between the latitudes of Stonehenge and Avebury there are 440 units of 207.36 ft, and in the direct distance between those two monuments there are 441. Thus the ratio between the two lines is 440:441, and that is the ratio between the earth's polar and mean radii.

Projected northwards from Avebury, this line continues to the 52nd parallel of latitude over a distance of 39.586909 miles, a hundredth part of the canonical measure of the earth's mean radius in ancient geodesy.

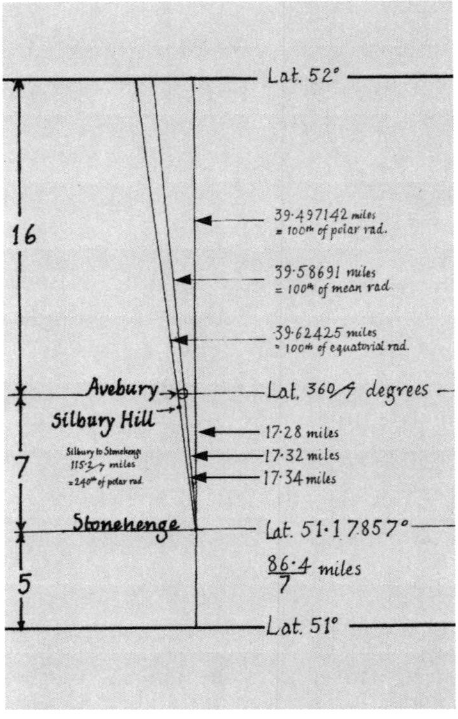

The third of the lines northward from Stonehenge runs at an angle of about 5 degrees west of north. Its calculated length, up to the latitude of 360/7 degrees, is 17.34 miles, which is 17.28 miles x 289/288, or 17.28 miles plus 316.8 feet.

The continuation of this third line, from the 360/7 degrees latitude of Avebury to the line of 52 degrees latitude, is 39.62425 miles, a hundredth part of the equatorial radius.

Figure 10.1 The three lines from Stonehenge to the 52nd parallel of latitude encode the polar, mean and equatorial radii of the earth.

The Stonehenge-Avebury line

The direct line from Stonehenge to Avebury is the most clearly marked, and its divisions are of great interest. There is all kinds of information in it - geodetic, metrological, musical and, at the root of all, numerical. There is also the pleasure and insight that comes from following this line on the ground and visiting the ancient sites that mark its course. As the evident base-line of a widespread system of augury, geomancy or mystical surveying, it sheds light on the origins of traditional sanctuaries in southern Britain - including the venerable site of Glastonbury Abbey.

At right angles to the Stonehenge-Avebury line is the line of the 'Roman road' that runs for about 14 miles dead straight from near Bath in the west and is orientated upon Silbury Hill. Its line is preserved by the Saxon boundary and highway, the Wansdyke, and by existing boundaries of county divisions. As it approaches Silbury, the road-line deviates slightly to the south and towards the southern flank of the hill.

This is undoubtedly a road once used by the Romans, for it passes through the Roman town of Velucio and continues eastward, though no longer on the same line, to the Roman station at Speen in Berkshire. Yet, like many such roads, it fits into the prehistoric landscape as if it were an ancient line which the Romans adopted. Its overall bearing, at right angle to the Stonehenge-Avebury axis and parallel to the Stonehenge-Glastonbury line, indicates a pre-Roman origin.

Also at right angles to the Stonehenge-Avebury line is the West Kennet long barrow. The axis line down its length extends westward, parallel to the straight line of Wansdyke, to Morgan's Hill where it meets a deviation of the Wansdyke. Other straight lines in this area, marking old boundaries and field-divisions, run parallel to the main Stonehenge-Avebury line. A striking example is the straight stretch of track between fields, now providing public access to the West Kennet barrow. It lies directly upon the main line and runs to the most prominent marker on its route, the great standing stone at the barrow's eastern entrance.

When looked at in detail, the Stonehenge-Avebury line reveals some peculiar features, unseen in any previous study of this subject. It is divided into sections by the markers along its route, and these are spaced at intervals that suggest a musical scale. But it is not as simple as that. There is not just one code of numbers in the divisions of this line, but at least two and probably more. You can read up the scale, from Stonehenge to Avebury, or in the other direction, and the scale varies. Moreover, in accordance with its 'up and down the scale' appearance, this line is divided midway into two halves.

The total length of this line is calculated to be 17.31927273 miles, *(this figure being virtually root 300)* From a spot within the north-east section of the Avebury enclosure, alongside the 'Cove' and at latitude 360/7 degrees, the line runs southward to the standing stone at the east end of the West Kennet long barrow. The distance is 7,464.96 or 864 x 8.64 feet. This is 27 times the radius of Silbury Hill and one part in 2,800 of the earth's mean radius.

South of the long barrow the line passes through a gap in the Wansdyke, alongside a prominent mound on Milk Hill and continues to the most beautiful spot on its route, the ancient enclosure containing the church at Alton Priors. Beside it grows a yew tree, split and twisted with age, and nearby fresh springs

Figure 10.2 Alton Priors church. Built on a mound with very ancient yew trees, a large sarsen stone with a hole bored into it may be discovered under a trap door in the nave.

rise up from a marshy bed. Here can be seen and felt the presence of a natural sanctuary. Inside the church is a rare reward for investigators of this line. A wooden trap door in the floor of the nave can be pulled open to reveal a great sarsen stone, over which the church was built. This stone has a large regular hole bored into it and is surely the original mark-stone, a surveyor's point on the Stonehenge-Avebury axis. It stands 25,920 feet from the start of the line at Avebury.

The southern half of the line, between the half-way point and Stonehenge, is sparsely inhabited, with few landmarks and largely controlled by the military. The one important structure situated on the line is Casterley Camp, an Iron Age enclosure of about 64 acres. Excavations there have provided evidence of an earlier sanctuary and settlement. The local village of Upavon is said once to have stood within its earth ramparts.

The southern edge of Casterley Camp is 36,000 feet from Stonehenge. At the same distance down the line from Avebury is a spot by the road at Bottlesford, and 2,520 feet further on is another roadside spot at Hilcot. This distance is the same as that between the opposite banks, north and south, of Casterley camp, measured along the line. What is more, the Hilcot point and the north edge of Casterley are equidistant from the centrepoint of the whole line - 7,200 feet. Around Hilcot and Bottlesford lines of roads and boundaries join together to form a mirror-image of the Casterley enclosure. The schematic map, figure. 10.3 *(opposite)*, shows the position.

These are the main intervals in the line between Stonehenge and Avebury: The measures are in feet.

Stonehenge - Casterley Camp = 36,000 ft

Avebury - Bottlesford = 36,000 ft

Casterley north bank - Hilcot south bank = 14,400 ft

These intervals add up to 86,400 feet, the same as the distance from Stonehenge to the base of Silbury Hill. The theoretical length of the line, 91,445.76 feet, is completed by two distances, the length of Casterley Camp and between Hilcot and Bottlesford, each estimated at 2,522.88 feet.

The 'mirror' aspect of the line is emphasized by two points along it, marking the 7-mile distance from Avebury and Stonehenge respectively. The first point, within the Bottlesford-Hilcot section, is now marked by a signpost at a meeting of footpaths; the second was a stone set in a wall (shown on old maps but since destroyed) within Casterley Camp.

There is no apparent mark at the centre of the Stonehenge-Avebury line. If ever there was one, it has long gone the same way as the many other small pointers that stood like milestones at intervals along the line. Apart from the buried sarsen stone at Alton Priors church, the only certain mark-stone that survives is the huge, erect sarsen that once blocked the entrance to the West Kennet long barrow.

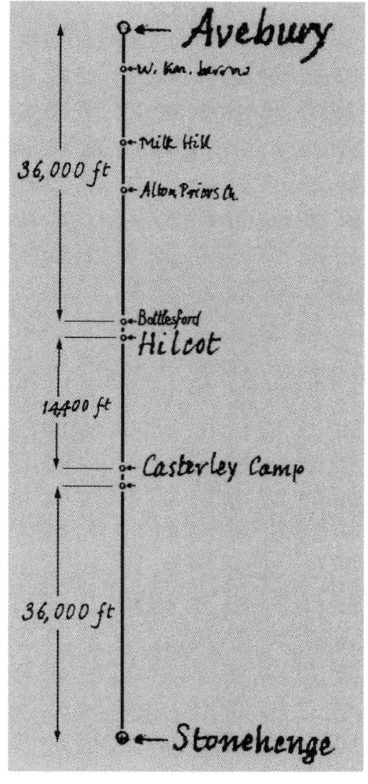

Figure 10.3 The principal divisions of the Stonehenge - Avebury line.

The correspondences between the northern and southern halves of this line, making each a reflection of the other, suggests first of all a musical notation. Musicologists can recognize the intervals and also the numbers involved, such as the octave 864: 1728 that Ernest McClain identifies in his works on ancient musical theory. At the same time, there are other likely interpretations, each involving number and measure.

Silbury Hill and the third, equatorial line

The great landmark on this line is Silbury Hill. Dated by archaeologists to around 2700 BC, this conical shaped mound, built up of solid chalk blocks and covered in turf, is spread over about $5^1/_2$ acres and measures nearly a third of a mile round its base. It is a huge monument, a fit tomb for an ancient Wessex emperor, but excavations into its core have revealed no trace of any burial.

The principal dimensions of Silbury Hill, adapted from the published figures, are (in feet):

> diameter of base 552.96; radius 276.48; height 129.6; slope 259.2, diameter of top 103.68 ft. The angle of slope is 30 degrees.

Figure 10.4 Silbury Hill, from the north *(photograph by kind permission of Lucy Pringle)*.

These dimensions are all duodecimal, significant multiples of 12, and are measured by the corresponding units, such as the common Greek foot of 1.0368 ft (= 12 x 12 x 12 x 12/20,000) or the grand old royal cubit of Egypt, whose length, 1.728 ft, is easily remembered because 1,728 = 12 x 12 x 12. This unit was applied to centres of national ritual, including the temple at Jerusalem and the Avebury - Silbury precinct. Measured by this cubit of 1.728 ft, the dimensions of Silbury are: base diameter 320, radius 160, height 75, slope 150, diameter of top 60.

Distances in the alignment system between Silbury, Avebury and other monuments can also be expressed by the cubit of 1.728 ft. From the centre of Silbury to the 'significant point' at Avebury is 2,880 cubits or 18 times the radius of Silbury. These measures are all geodetic, simple fractions of 20,901,888 ft the canonical length of the earth's mean radius. To obtain this, the radius of Silbury is multiplied by 75,600 and the distance from Silbury to Avebury by 4,200.

The diameter of Silbury Hill, multiplied by $^{22}/_{7}$, gives the measure round its circular base as 1,737.874286 ft. That is 500 units of a familiar measure, the royal yard (3.4757486 ft) that defines the Stonehenge lintel ring and represents a 6-millionth part of the polar radius.

The distance from Stonehenge to the nearest, southern foot of Silbury Hill is 86400 ft, or 50,000 royal cubits of 1.728 ft. From there the line continues to latitude 360/7 a further distance of 5,184 ft or 3000 cubits. That makes its total distance, from Stonehenge through Silbury Hill to latitude $^{360}/_{7}$, equal to 53,000 cubits or 91,584 ft. That is 265 to 264 in relation to the first, meridian line.

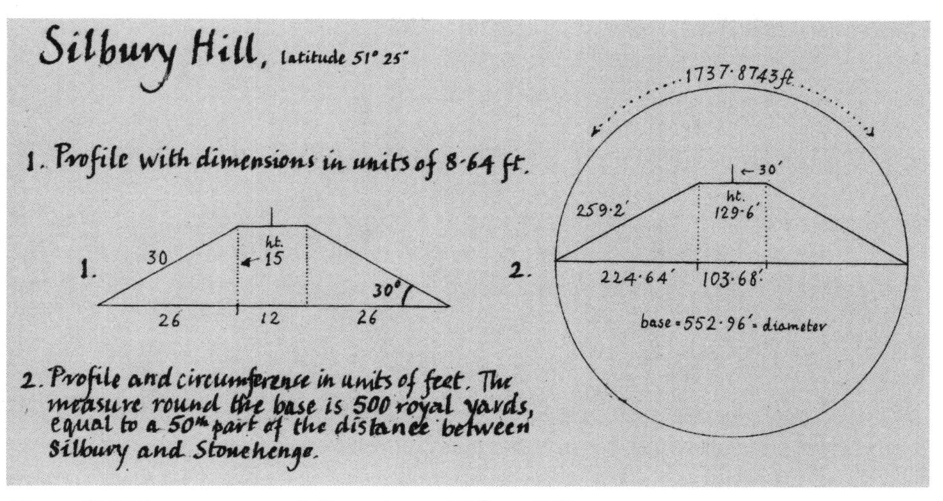

Figure 10.5 The geometry and dimensions of Silbury Hill.

This is a surprise result. It has previously been supposed the the ratio between the polar and equatorial radii is 288 to 289. Yet the evidence here suggests that 264 to 265 was the ratio used by the ancient surveyors. This means that the indicated length of the equatorial radius is 3964.675325 miles.

The Woodhenge-Gare Hill alignment

Relating to the Stonehenge-Silbury line, as parallels or perpendiculars to it, are other lines across and beyond Salisbury Plain. Best marked is the alignment at right angles to it that that takes in some of the major monuments in the Stonehenge area. They include Woodhenge, situated 10,000 ft to the north-east of Stonehenge, and the Cursus, the linear earthwork about 1.7 miles long that runs to the west of Woodhenge.

This alignment starts at the centre of Woodhenge and is directed towards the chapel of St Michael on Gare Hill. This dedication to the Archangel is ancient, and so is the site itself. Other alignments and old trackways converge upon it. Gare Hill stands upon the direct line between Glastonbury Abbey and Stonehenge. It is situated upon the county boundary which is also the old boundary between Wessex and the West Country. There was never a parish church on Gare Hill but a 'chapel of ease', a resort for travellers and pilgrims. The present chapel, now a private residence, replaced the medieval building on the site where, earlier perhaps, was a hermit's cell and oratory.

The prominence of Gare Hill in the ancient alignment system suggests that the original structure upon it was some kind of surveyor's mark.

Figure 10.6 The former church of St Michael on Gare Hill occupies an important ancient site upon the Wiltshire-Somerset border.

Whoever was stationed there - priest, Druid, hermit or inspector of lines and trackways - maintained a beacon light for the guidance of wayfarers.

The total length of the Woodhenge-Gare Hill alignment is 23.04 miles, equal to 24 Greek miles. Its eastern section is marked by aligned monuments and, along the ground, by the linear earthwork that forms the northern edge of the Cursus.

Woodhenge at the eastern end of the alignment was discovered and excavated in the 1920s. All that remains to be seen of it are the concrete slabs, placed by archaeologists to mark the holes that once held large wooden pillars. The pillars were arranged in concentric ovals and probably supported a thatched roof. Their careful, geometrical planning indicated that Woodhenge was a temple, a place of ritual and science. The date allotted to it is about 2,300 BC.

The alignment west from Woodhenge was described by John North in his great treatise on megalithic science, *Stonehenge* (1996). It begins at the centre

Figure 10.7 Woodhenge. This strange collection of concrete stumps marks several ovals of concentric post holes, all that can be seen of what was once an impressively large neolithic roundhouse, egg-shaped and with its axis and main entrance aligned to the midsummer sunrise.

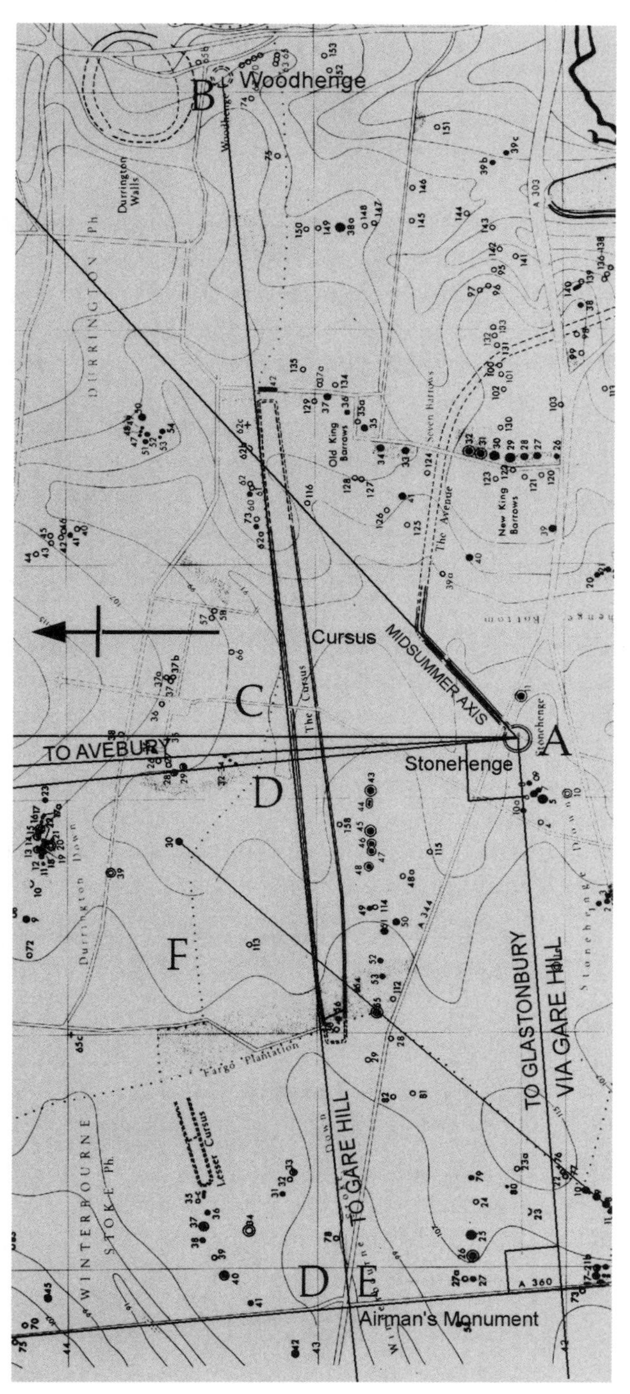

Fig 10.8 The straight line from the centre of Woodhenge to Gare Hill, 23.04 miles to the west, passes over the Cuckoo stone and along the northern edge of the Cursus. Its next mark is an ancient crossroads, where it forms a right angle with the north-south road. This former trackway is parallel to the Stonehenge-Avebury line, and 1.44 miles distant from it.
Key: A. Stonehenge, B. Woodhenge, C. The Cursus, D.D. Right angles, E. Airman's Cross, F. ancient boundary parallel to Cursus line. Also shown are the lines to Glastonbury and along the Stonehenge Avenue to Goring.
The arrowed line points due north.

of the monument and its line is marked by the site of a former wooden post a few yards to the west. About 1,700 ft beyond is a bulky sarsen block, known as the Cuckoo or Cuckold stone, and the line then passes by the north end of a long barrow, a point where, as North shows in his book, eight alignments between other local long barrows cross each other.

Then comes the long stretch where the line runs along the northern bank or ditch of the Cursus. The purpose of this long-distance earthwork, and of the labour and skills that went into its construction in about 3,400 BC, is one of the great archaeological mysteries. Its name comes from the old antiquarian theory that it was an ancient racecourse. Modern researchers look at it (and at similar monuments elsewhere) from an astronomical point of view; and it seems also to be an instrument of the old surveyors. But there is still no practical explanation why its builders, around 3,400 BC, devoted years of effort to planning and making it.

The last point on this section of the Woodhenge-Gare Hill line is one of the oldest features of Salisbury Plain, a crossing of roads at a spot called Airman's Cross because of a monument that now marks it. This is a crucial spot in this investigation. The crossroads is situated on a long, straight road, overlying an ancient trackway, that runs precisely at right-angles to the Woodhenge-Gare Hill alignment. This alignment is itself at right-angles to the line from Stonehenge to Silbury Hill and beyond. That means that the road approximately north-south through the Airman's cross is parallel to the Stonehenge-Silbury line. And the distance between them is significant - 1.44 miles or 1.5 Greek miles. This is a sixteenth part of the complete Woodhenge-Gare Hill alignment. Fig 10.7 shows the features in the Stonehenge landscape that run parallel or at right-angles to the Stonehenge-Silbury line.

The Circle of Perpetual Choirs

Westward from Stonehenge runs the line to the site of the old church at Glastonbury Abbey - the traditional site of the first Christian foundation in England. The length of this line is 204,120 ft (38.66 miles). It is divided into two parts, relating as 11 to 9, by the site of St Michael's chapel at Gare Hill on the Wiltshire-Somerset border.

This line forms one side of the great decagon whose points mark the sites of the ancient Perpetual Choirs. Three of them are mentioned in Welsh bardic records - at Stonehenge, Glastonbury and Llantwit major in South Wales, the famous Celtic college of St Illtydd, with a mystical foundation legend similar to Glastonbury's. A fourth, north-east from Stonehenge, is at Goring-on-

Thames. Another site, further round the circle, is Croft Hill outside Leicester, a spot where open-air assemblies were held into historical times. Paul Devereux identifies Croft Hill as a central site of old England. Lines between these sites make angles of 144 degrees, the angle of a regular ten-sided polygon.

The central point of the figure is at Whiteleafed Oak, a spot where the three counties of Worcester, Hereford and Gloucester have their meeting-place. It is an apt coincidence that the cathedral cities of these three counties hold an annual Three Choirs festival. The legends of Whiteleafed Oak identify it as a former Druid grove with the sacred oak at its centre. Today it is an obscure, unmarked spot, hidden among the most ancient hills of England, but geographically and atmospherically it has the attributes of a natural sanctuary.

Perpetual choirs were a Celtic institution, from pagan into early Christian times. In Iolo Morganwg's *Triads of Britain*, translated from Welsh, it is stated that 'in each of these three choirs there were 24,000 saints; that is, there were a hundred for every hour of the day and the night in rotation, perpetuating the praise and service of God without rest or intermission.'

The function of these choirs was to maintain the enchantment of Britain - by chanting. Their song was a constant religious chant, varying with the cycles and seasons and, like time itself, never ending. It was reflected in popular mode by the music that was heard at festivals around the country. In that way, everyone was held under the same musical spell that maintained harmony in their relationships and surroundings.

The line from Glastonbury makes an angle of 144 degrees at Stonehenge, and continues thence to the next choir site 204,120 ft further on. This takes it to a spot on the Thames at Goring which, as gor-ting, means a meeting-place of choirs. The actual spot is on the east bank of the river at Cleeve Wharf, another spot with a resonance of sanctity. A house there is appropriately called the Temple. From Stonehenge the line to Goring passes directly down the Avenue and is orientated to the point of sunrise at the summer solstice.

The dimensions of the Perpetual Choir circle are of great beauty and interest. The principal unit in the scheme is indicated by an obscure and controversial monument located on the summer solstice line from Stonehenge to Goring. Named Peter's mound after C.A. 'Peter' Newham who discovered it, it is a small earthwork visible on the horizon from Stonehenge and marking the point of sunrise on the longest day. Alexander Thom made a careful survey of it in 1978, and established its distance from the centre of Stonehenge, 8,981 ft. Adjusted by about three inches to 8,981.28 ft, this distance is equal to 4,400 units of 2.0412 ft. A hundred thousand of the latter make 204,120 ft, the distance from Stonehenge to Glastonbury's

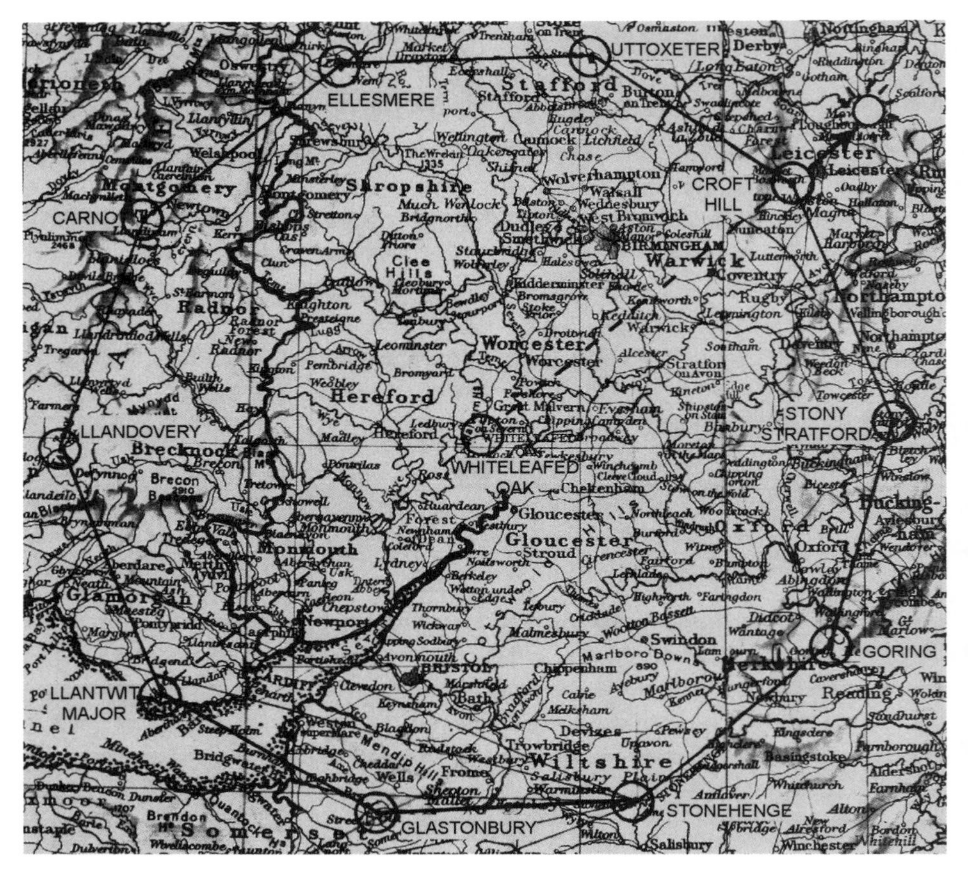

Figure 10.9 The Ten Sites of the Decagon. The five sites clockwise from Goring to Llandovery have enjoyed a long history of monastic settlement and learning. The line running through the decagon from Llantwit Major to Croft Hill near Leicester, follows the path of midsummer sunrise and is parallel to the Stonehenge midsummer line along the Avenue towards Goring on Thames. *(The theoretical coordinates of each point on the decagon are listed opposite).*

old church or, in the other direction, from Stonehenge to Goring. The significance of this measure is that 20,412 ft, multiplied by 1024, or 2^{10}, is 20,901,888 ft, the canonical figure for the earth's mean radius.

Peter's mound was investigated by archaeologists who dismissed it as a modern structure. That has since been disputed. Its key position and metrological relationship to Stonehenge and the perpetual choir circle give evidence of its basic antiquity.

To find the radius of the perpetual choir circle, the side of the decagon, 204,120 ft, is multiplied by the golden section number, 1.618033989... Like all 'irrational' proportions (*pi*, root 2, root 3, etc.) this was made rational by two whole numbers, in this case 160/99 = 1.61616. That makes the radius of the circle 329,890.91 ft or 10!/11 ft. Multiplying that figure by 44/7 gives the circumference of the perpetual choir circle as as 2,073,600 ft or 100 x 12 x 12 x 12 x 12 ft.

The ratio between the circumference of the perpetual choir circle and the perimeter of its inscribed decagon is 2,073,600 to 2,041,200 or 64 to 63.

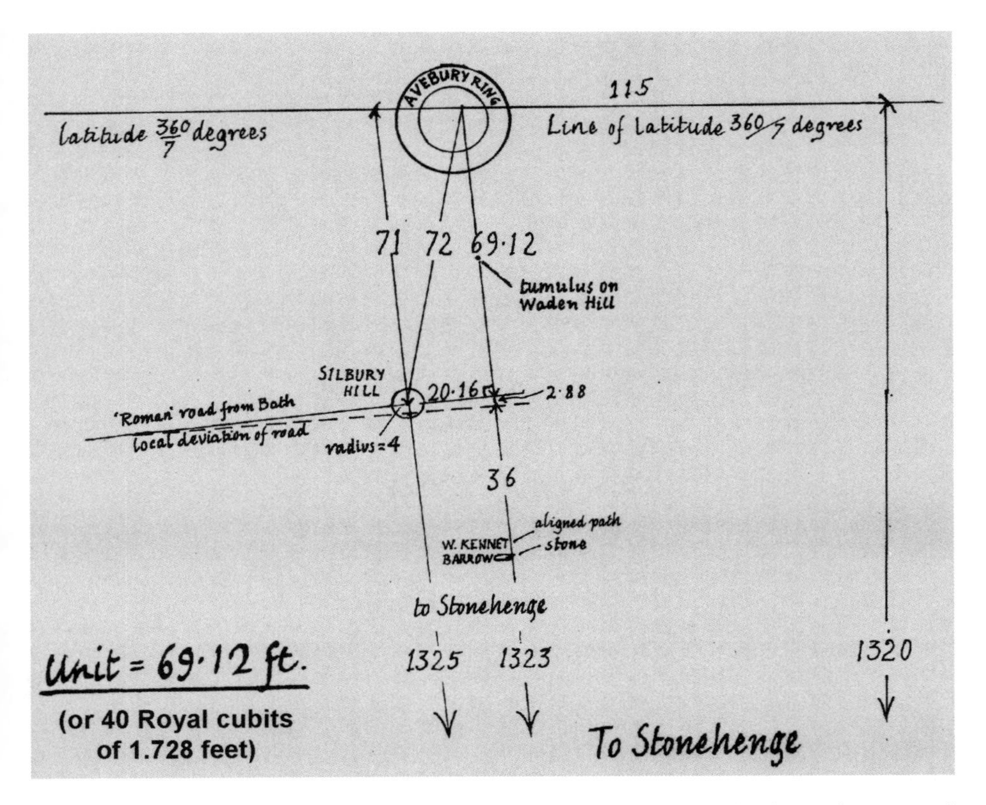

Figure 10.10 The distances between Silbury Hill and other prehistoric landscape features and monuments on the three identified meridian lines between Stonehenge and Avebury, in units of 69.12 feet. The respective lengths of these lines are 1320, 1323 and 1325 of this unit. Projected northward to latitude 52 degrees these lines, in the same unit, are numerically equal to the dimensions of the earth's polar, mean and equatorial radii.

(Note that 69.12 feet is 40 Royal cubits of 1.728 feet; one degree of latitude averages at 69.12 miles)

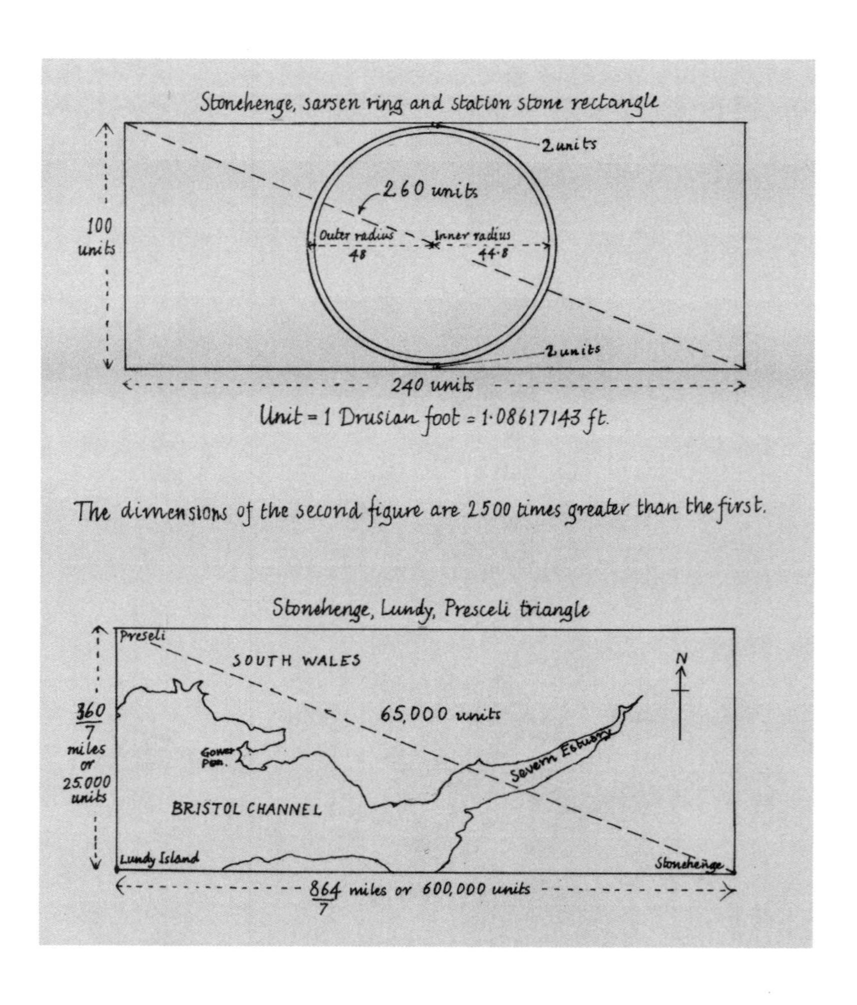

Figure 11.1 Illustrated here is the relationship between the Stonehenge station stone rectangle and the Lundy-Stonehenge rectangle. The common unit throughout is the Drusian foot, 1.08617143 feet, which is two fifths of the 'Astronomical megalithic yard' (AMY). The scaling between the two rectangles is 1 : 2,500.

Chapter Eleven

Dates and Speculations

Britain is the most archaeologized country in the world. Yet after centuries of digging and research, the nature of prehistoric culture is still a mystery. Ancient artefacts say nothing for themselves, so their interpretation depends on whatever myths and theories of prehistory happen to prevail. These in modern times have been conditioned by scientific materialism and rooted in the 'rise of man' overview of Darwinism.

The effect of this has been divert attention from the evidence of astronomy, surveying and other sciences in prehistoric Britain. Even when the evidence is considered, the conditioned view is that ancient science was a primitive affair. The tribal folk who practised it were stumbling from primal ignorance, taking first steps towards the light of modern understanding.

This evolutionary and, in effect, racist attitude towards intelligent humans of a different age is now on the wane, so there is no point in dwelling on it. It is easy to criticize, but if the established overview on ancient Britain is totally wrong, what sort of picture should replace it?

The picture that emerges from this book is of an ancient British culture that combined simple living with an informed and comprehensive code of science. It was a very different science from the modern, pragmatic version, for it was based on traditional knowledge that was said to have been divinely revealed. It recognized the spiritual powers in nature and the immortality of the soul.

At the beginning of the historical period this religious and ritualized science was still practised in the remoter parts of Europe, particularly in the British Isles. The old Druid colleges were famous throughout the Continent for their learned teachers and the education which led to initiation into the Mysteries. In the last days of Druidry, shortly before its reformation that produced the native Celtic Church, Julius Caesar wrote a respectful account of Druidic science and wisdom.

He said that the Druids used the Greek alphabet for their accounts and secular matters, but for instruction in sacred subjects they would not allow writing, insisting on everything being taught and learnt by heart. One reason for that was that writing destroys memory; another was that anything written can fall into the hands of persons unqualified to read it.

On the subject of Druidic studies Caesar wrote:

> *'The cardinal doctrine which they seek to teach is that souls do not die, but after death pass from one to another... Besides this, they have many discussions concerning the stars and their movement, the size of the universe and of the earth, the order of nature, the strength and powers of the immortal gods'.*

The Druids of Caesar's time lived two thousand years after the Stonehenge builders and two thousand years before us. But in the first period, before the rise of cities and the cult of progress, time was not such a barrier between successive generations living similar lives in the same country, century after century. There was continuity, on all levels, in the rituals and customs of daily life and in the sacred tradition of science and philosophy that the Celtic Druids inherited from ancient predecessors.

Here, finally, come the questions that we have avoided throughout this book. When and where is the origin of this tradition? What is the date of the earliest pattern of surveyed lines and centres across Britain?

These questions have been avoided because we can not really answer them. The overall pattern of ancient surveying is clearly a work of inspiration. It began at some unknown time and was renewed and added to over the following millennia. All that can be attempted is to trace the tradition backwards, from the Iron age and the time of Caesar's Druids until it vanishes into the mists of pre-pre-history.

The Stonehenge experts say that by the time Caesar came to Britain, the old temple was abandoned and probably in ruins. Yet the tradition of great works by surveyors in the past was still alive, for Caesar was told about the triangle that outlines Britain, complete with its dimensions. Also remembered were the lines that ran from coast to coast and from town to town across the island. Along sections of them were paths and paved highways from times beyond memory. Local people explained them through legends - telling, no doubt, of gods who first made them and great rulers who repaired them and confirmed their sanctity. The literate minority - Druids and their initiates - retained some knowledge of the ancient science, for in Caesar's record one of their skills was geodesy - surveying and measuring the earth.

The Great Paradox of Prehistory

The most obvious suspect as originator of the great survey of Britain is the Bronze age Wessex culture, centred upon Stonehenge and Avebury. It was in every sense a golden age, with all the attributes that the alchemical philosophy classifies under gold - divine kingship, the inspired law-giver, the heroic warrior and a social hierarchy, reflecting the revealed order of the heavens. In this form of society - which has been known at different times throughout the world - the people are divided into twelve tribes, in imitation of the rational order of the zodiac. In each section is located an episode in the official myth which describes the adventures of a solar hero in his yearly course round heavens. And each section has its note in the twelve-note scale of music which perpetual choirs maintained, day and night, in the monastic colleges.

It was a spell-bound society, held under a priestly enchantment, highly ritualized and never changing from one generation to another. Its values were solar and rational, and its religion was bound up with the rational sciences, astronomy and surveying. In connection with these and other sciences, Bronze age monuments in earth and stone imposed a new pattern on the surface of Britain. In that period, around 2200 BC and the dawning of the age of Aries, Stonehenge was completed, the great stone circle at Avebury was erected and a tall sarsen stone was set up at the east end of the West Kennet long barrow. This implies that the line itself, and the entire system of surveyed lines in the Wessex region and beyond, were laid out in the Bronze age.

But that cannot be so. The sites of Stonehenge and Avebury were known and sanctified long before the Bronze age came to Wessex. The West Kennet barrow, at right angles to the line between those sites, was built more than a thousand years earlier. The Bronze age people, it now seems, repaired and made permanent a system which, even in their time, was archaic and venerable.

This is where the trail fades away. Before the Bronze age, in the period called Neolithic, archaeology tells of hunters and primitive farmers, with a local, subsistence economy, and with no apparent need or resources to conduct large-scale surveys of Britain. Yet the evidence set out in this book is of an archaic culture with an elaborate code of science, based on the magic of number.

There is no reconciling this evidence with the current evolutionary view of prehistoric life and culture. Readers are invited to check it for themselves and, if it holds up, to adapt their thinking accordingly. A possible view, requiring nothing new in the way of faith or fantasy, is the traditional one - that our ancient ancestors were like the mythical giants and heroes, nearer to the gods than we are, and capable of such great achievements in science and ritual magic that we are only just beginning to recognize them.

- End of Book Two -

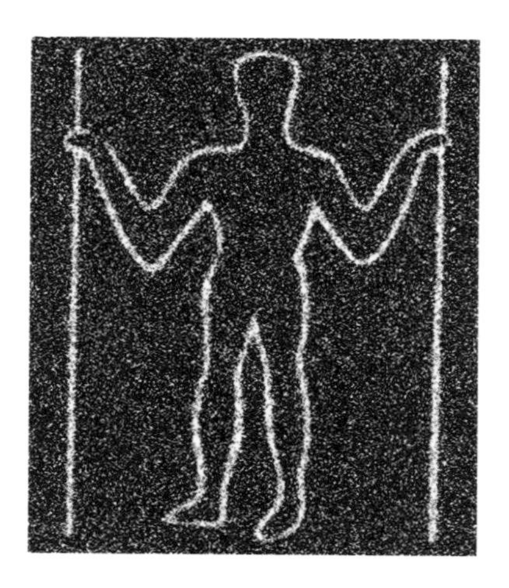

- APPENDICES -

APPENDIX ONE

A METROLOGICAL ASSESSMENT OF THE SYSTEM REVEALED IN *THE MEASURE OF ALBION*

by John Neal, Principal of *The Secret Academy* and author of *All Done with Mirrors.*

If Heath is the mathematician-astronomer and Michell the antiquarian-philosopher, then my contribution to this work has been that of the bricklayer, in cementing things together. Through the revealed metrology that is of such importance in supporting the radical ideas presented in this work I will attempt to explain that not only are their findings plausible, but also are much more than probable.

If I may be regarded as something of an authority on the subject of ancient metrology it is only because it has almost entirely faded from academic or public interest. The whole stunning subject, once the darling of such august bodies as the Royal Society and the French Academy of Sciences, where it was an issue of regular debate, is now virtually dead. The reasons that the subject has fallen from the curriculum of the science of history or the history of science, is that the enigmas that were deliberated by the early metrologists were never resolved. Consequently, the study came to be regarded as something of a scholarly nightmare.

At first glance, this would seem to be the case. The sheer variety of names for the measures - leagues, miles, schoenus, atur, parasang, furlongs, stadia, rods, poles, reeds and perticae; the lesser modules of fathoms, paces, steps, cubits, bracciae, remen, pygons, pygmys and feet; their sub multiples of palms, hands, digits, inches, su-si's, just to mention some of the European and Middle Eastern terminology, is intimidating. Add to this the acknowledged variations exhibited by each of the established modules and the whole subject appears random, confused and quite disconnected.

Although I had taken rather more than a passing interest in ancient metrology for many years, it was not until I tackled the subject intensively and systematically that its underlying elegance and simplicity became apparent. This initially came about through simplification. By ignoring all of the multiples and subdivisions and concentrating upon the variety of known basic foot measures and then comparing them with each other, that progress in understanding the system of metrology was accomplished.

THE IMPORTANCE OF THE FOOT

The foot is regularly considered to be equivalent to the human appendage of that name, but none of the values of the ancient feet favourably compare, in a literal sense, with the size of a human foot. The very term 'foot', as a length, should therefore be considered in another context, such as pedestal, or basis, or *root*. All nations, eventually, based their measurement systems upon one or the other of the feet, and there was not a very large difference in their particular lengths. There are probably only 12 distinct feet from which all other measures (modules) are derived.

In the nations of the ancient world all of the various feet were used concurrently, we merely term these feet Roman, Greek, or Egyptian etc. because the various national bureaucracies adopted one of them for regulation within their society. All of the others continued to be used by the artisans because they comprise a single system, the purpose of which was the maintenance of rational integers within their designs. The builders of the ancient temples and surveyors of the road systems used these measures to astonishing degrees of accuracy, which is how we are now able to clearly identify their selected modules.

SOME KEY POINTS IN ANCIENT METROLOGY

Within all cultures the individual feet, and all other modules proportionately, display a considerable variation. This has been commonly regarded as slackness in the maintenance of standards, but this cannot be the case because these variations are found to be identical in ancient societies.

It was Flinders Petrie who first remarked that measures regularly varied by the 170th and 450th part, but these figures are now able to be refined to the 175th and 440th parts. It was John Michell who perceptively identified the separation of the 175th part between measures, not only identifying this difference in the modules of the royal and common Egyptian values, the Greek, Roman and sacred Jewish, but also discovering the absolute values for these modules.

The identification of these absolutes had previously been a bone of contention between metrologists. Without them, approximations could be used and indeed were used to support just about any leaky theory because the spread of values allowed so much latitude. Only when one has identified the absolute values can the precision of the ancient engineers be appreciated, and a theory rigorously checked and firmly substantiated. For example, the miniscule part of the 440th would appear to make no difference at the length of the foot, but it is a full seven inches in the length of the Parthenon. At geographic, or itinerary distances, this meticulousness may be shown to go to astounding levels of perfection.

I was able to identify the 440th separation by comparing the values accurately identified by the great metrologist, Livio Stecchini, for the Roman and Greek feet, with those absolute values put forward by Michell and by noticing that the Royal cubit of the Pyramid, identified by Petrie as 1.71818 ft, differed as 441 to 440 to a Royal cubit expressible as $^{12}/_{7}$ English feet, - the 'root' Royal cubit.

These two fractions, the 175th and the 440th, have practical mathematical application in the maintenance of integers in circular constructions. $^{22}/_{7}$ was the most commonly used approximation for pi in the ancient world - it is the simplest rational fraction to approximate the irrational pi, being accurate to 99.96%, or two feet in a mile. If the diameters of circles are multiples of seven, then the perimeter unit will be integral with the diameter unit. If they are not multiples of seven then the perimeters become rational by using a module that is the 175th part greater to measure the resulting perimeter. For example, if the diameter is four units then the perimeter will be 12.5 in a module the 175th part greater, instead of the fractured number 12.57142857 obtained by using the same unit to measure the perimeter as the diameter.

The fraction of the 440th part has a similar function, that of maintaining integers through differing modules. The best example of this is to take the canonical number 360 as a circumference, in units of English feet. The diameter measures 114.545454.. English feet. But this is exactly one hundred royal Egyptian feet of what I have termed the Standard classification. The royal Egyptian foot is exactly one and one seventh English feet, at what I have termed its Root value. The number converts to Standard by the addition of the 440th part. Michell showed, in *The Dimensions of Paradise*, that the fraction $^{441}/_{440}$ converts the mean radius of the earth to the polar radius using the values used by ancient geodesy.

Metrology is seen to derive from the behaviour of number itself. Of particular interest to this book is that a diameter of 100 Common Greek feet of the Standard Geographic classification is 104.27245 English feet, the outer diameter of the lintel circle at Stonehenge, which as a consequence has a perimeter of 360 Mycenaen feet, identified by Stecchini. This is the 440th part less, at Root Geographic value, than the diameter measures, which are all Standard Geographic values.

This Mycenaean foot, at 0.910315ft, (variants of which are termed Italic, Oscan and Assyrian), fits the Meridian degree at around latitude 38^{o}(or 3/7ths of the distance from equator to pole) an exact 400,000 times, because it is a Root Geographic value. This is a particularly interesting degree of latitude, not only because it embraces the Aegean but also because 360 degrees of this length would be the perimeter of a circle, the diameter of which would be the polar diameter, exactly as given by Michell's radius of $3949\,^{5}/_{7}$ miles.

The Mycenaean foot is also exactly 9 to 10 of the Root Geographic Greek foot of 1.01146122ft, also identified by Stecchini. This was the value that was pivotal in deciphering the metrological system as a whole, because regarded as a simple number it is $(^{176}/_{175})^{2}$. Having realised this fact, it was obvious to me to search for a Greek foot intermediate between 1.0114612 and the English foot. This should have a length of 1.0057142ft, (the decimal expression of $^{176}/_{175}$). Martin Folkes had identified a very good example of this very value, as reported to the Royal Society in 1736. He found it with five other modules engraved on an official standards stone at the Capitol in Rome.

This firmly established the English foot as one in the series of the known values of the Greek feet. Both of these values 1.0057142ft and 1.01146122ft are respectively 440 to 441 of the values identified by Michell as Greek feet at 1.008ft and 1.01376ft.

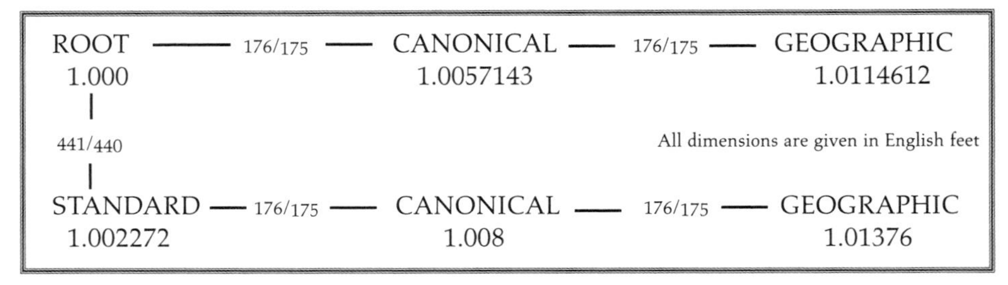

From the Root of one English foot the values must be tabulated in two rows, the fraction linking each of the variations across the rows is $^{176}/_{175}$, and each of the values in the top row is linked to the value directly below as $^{440}/_{441}$. 'Root' prefixes the descriptive terminology from Least (smallest value of the measure) to 'Geographic' in the top row and 'Standard' in the bottom row. For example, 1.008 feet is Standard Canonical and 1.0114612 is Root Geographic etc. As well as these values being actual known measures, they may also be regarded as the formulae by which any other module is classified. This is far from being the full picture, but this simple arrangement is sufficient to illustrate the principles here.

There is now a method of identifying and classifying any measure whatever, while previously this has always been lacking in the study of metrology, and has been the primary cause of much confusion. A coherent system has emerged from the disorder of metrology, one first-fruit being that this system has enabled the first scientific account of prehistoric surveying in Britain to be undertaken by Heath and Michell.

THE ITINERARIES

Even among the educated classes and archaeologists in particular, there is a prevailing belief that British history and civilization began with the Romans. The evidence would imply that they made sweeping social changes, but then so did the Normans, and the Roman intrusion should be viewed as just another order imposed for a brief period of our true historical span. Although the Romans introduced bricks and mortar in the construction of their towns, towns had been established, using the same principles, for as long can be imagined.

These principles were siting, delineation, orientation and consecration, then the Cardo, or north-south baseline was laid and the ground was quartered. The planning experts who attended to these secretive rites were working from extremely ancient and universal traditions. Similar methods were employed in Persia, India, China and Mesopotamia. Architectural historians are unresolved as to whether the methods were of Mesopotamian or Celtic\Teutonic origin.

Caesar, in his Gallic Wars, records the towns of the Gauls as being connected by perfectly straight roads, the lines of which carried through the towns regardless. Although Britain is covered in a network of ancient 'Roman' roads, there is little evidence that the necessary surveys had been carried out by the Romans. Their primitive surveying instruments, such as the groma, was only of use over short distances. The Roman was never born who could have begun a road to link Exeter and Lincoln, for the simple reason that he would have had no idea which direction to take.

All of the methods, which have been tentatively forwarded as to how it could have been done, are unsatisfactory. From the trial and error of long lines of men constantly to-ing and fro-ing till they come to a semblance of a straight line, may be dismissed. So too can beacons of fire in baskets to aid visual sight lines, the curvature of the horizon precludes it as being of use over very long distances. It has even been suggested that homing pigeons were released and their flight paths noted. These difficulties are most ably documented by Hugh Davies in his *Roads In Roman Britain*, and along with all other authors on the subject, he has had to admit defeat as to a rational explanation.

The only known reference to a method, which is feasible, was recorded by Heron of Alexandria, who proposed that an accurate base line be measured and the area be mapped from this line by triangulation. This, of course, is the only known way to do it, but the Romans never used the method, because by Heron's time of the first century, the roads were already built. There is an example from Greece of a sighting device, called an alidade, placed upon a horizontal plate that is marked out in angular degrees. Add lenses to the sighting device and you have a theodolite. Evidence, as gathered by Robert Temple in his book *The Crystal Sun*, proves that lenses were abundant in the ancient world.

Everything was known at one time or another that would be sufficient to make accurate instruments to enable the necessary surveys to be undertaken. This was possible, indeed it was done, long before the Romans arrived in Britain. All they had to do was enquire of the locals which track went where, and how far, then simply had to resurface the existing trackways to accommodate their greater demands.

Roman roads invariably connect prehistoric sites of the earlier survey, as noted from Codrington's *Roman Roads in Britain*:

'Changes of direction from one straight line to another, when the change is not at a station or some other point through which the road had to pass, almost always occur at points on high ground. There are several instances where a barrow or tumulus was the landmark, the road passing round it or nearing it. Silbury affords one example, and Brinklow, on the Fosse, another.'. Like many other authors, he too goes on to speculate that where the distant points were not visible then intermediate high points would be marked to maintain the straightness of the survey; but this cannot be the case.

Many writers on the subject of Roman roads remark admiringly, on the one hand, how accurate they were, yet on the other hand, when it comes to the recorded distances between points connected by the roads, remark how imprecise they were. A careful analysis of these recorded itinerary distances provides as much confirmation for a pre-Roman origin of the system as does the physical evidence.

The principle Roman measurement of itinerary distance was the *mille passum,* literally one thousand paces, each of five feet. 5,000 thousand feet miles were used by many cultures; the mile of 5,000 English feet was commonly used in Britain until supplanted by statute of Elizabeth I for the mile of 5,280ft. Both these miles had been used in Britain since time immemorial, as an analysis of the Antonine Itinerary clearly shows. This document, through many copyists, has survived from Roman times; it gives the distances between Roman settlements throughout the areas of Britain that were occupied by them. Although certain errors must have crept in, overall it is remarkably accurate. No archaeologist would make this statement, because it is generally taken for granted that all of the distances were given in terms of the *mille passum* of 5,000 Roman feet. In these terms the errors appear enormous, but if the distances are carefully plotted on the map and the length divided by the number of 'miles' given in the Antonine Itinerary, other known lengths for the mile fit these distances with remarkable regularit and accuracy,, particularly the mile of 5,280ft.

The Itinerary comprises fifteen sections, each listing a number of connected towns or settlements. The first astonishment is delivered immediately in Itinerary I, which begins at Hadrian's Wall. From High Rochester, Northumberland, the distance given to Corbridge is 20 miles, but consultation of the map gives 23.04 statute miles and this is exactly 20 nautical miles of 6082.56 feet (the Admiralty value for Britain). The nautical mile is 6,000 Greek feet of 1.01376ft or 10 stadia, thus the distance is 200 stadia overall. (The geographic mile is also the measure used in the itinerary at Itinerary 13, from Canterbury to Dover)

Then from Corbrige to Ebchester, Durham, the Itinerary gives 9 miles, whereas the actual distance is 9.6 statute miles. Not only is this exactly 9 x 5,000ft miles that is a variation of the ancient British foot of the 'yard and full hand', but it is also 10 Greek miles of 5,000 feet each of 1.01376ft. It is not until the distance given from Catterick to Aldborough, stated to be 24 miles that we have a recognisable Roman mile of 5,000ft of 0.9732096ft.

Examples of the 5,000 English feet mile are exactly the 16 miles given between Wall, Staffs to Mancetter, Warks. Also the 17 miles from Aldborough to York. An example of the 5,280ft mile is given by the distance from Chester to Tilston, listed as 10 miles.

These points will not be laboured here, as my research goes on, but it is remarkably obvious that these Roman itineraries of Britain are quite deliberately contrived in terms of many more modules than the Roman mile. The pioneer of aerial archaeology, Jacques Dassié, is presently, quite independently, making similar findings in analysing the itinerary measures throughout Gaul.

As Heath and Michell have found the English mile to be of such importance to their own investigations, it would be as well to thoroughly investigate its composition at this juncture.

ENGLISH MILE, PERSIAN FOOT

The British measurement system is an accretion of elements from three quite separate bases. The factor which binds these elements together is their integrity of length with the English foot, which itself is a variant of the measures we term Greek. The poles, chains and furlongs were, what are termed, Saxon measure. At 16.5ft, 22 yards and 220 yards respectively, they are basically multiples of the Saxon foot of $^{11}/_{10}$ English feet thereby giving the origin of these as, more properly in Saxon feet, sexagesimal 15ft, 60ft and 600ft. These Saxon or Northern values share an origin with what we term Sumerian, in both their length and counting base.

The English mile, however, originates in yet another system, whose root we term Persian. If the English mile were taken as the traditional 5,000ft, which the term 'mile' implies, then it is composed of feet of 1.056ft, and the basis of Persian measure was the foot of Darius at exactly 1.05ft, and 1.05 to 1.056 is the familiar $^{175}/_{176}$.

The most interesting and conclusive example of a foot of this length comes from Greece. Allegorical stories often concern measures, a definition of the Greek stadium of 600ft was said to be the length that Hercules could run in a single breath. Moreover he accomplished this feat at the stadium in Olympia, thereby defining the length of the track. Plutarch says that Pythagoras ingeniously calculated the height of Hercules by comparing the length of various stadia in Greece. A stadium is 600 feet in length, but Hercules' stadium at Olympia was longer than the normal 600 Greek feet. As the start and finish lines are still clearly delineated there, it may be measured and is a tad over 193 metres - it is 600ft of 1.056ft. Therefore, 25 of Hercules stadia are identical to the old English league of three miles.

This league was also the principle itinerary distance used in Arabia, called a *farsang*, derived from the Persian *parasang*. The measurement standard adopted by the maturing Arabic Empire was the Hashimi cubit, given as 649mm. (at half this length it would be termed a Standard Geographic Persian foot, that is, 1.05ft x 1.01376). This was the measure adopted by Charlemagne as one of the cementing links between his Frankish Empire and that of the great Arabian league of states represented by the Caliph Harun-al-Rashid. It is widely believed that on the adoption, in 785 AD, of the Hashimi cubit by the Franks as two *pied de roi*, that this is when it first made its appearance in Europe. But Jacques Dassié has produced some very convincing proofs that the lineage of the *pied de roi* in France extends back into prehistory.

Just as the *Antonine Itinerary* is the British reference for Roman distances, in Gaul, it is the *Tables of Peutinger* that are consulted. This mediaeval document is a copy of a Roman map of the whole Empire, and the distances between settlements are given in terms of the 7,500ft, or $1^{1}/_{2}$ mile, league. Exactly as my own scrutiny of the British measurements has revealed that there are a variety of miles, recognisable by their accuracy in the Itinerary, so Dassié obtains identical results in France with the recorded lengths of the leagues, both from the *Tables of Peutinger* and from the surviving milliary columns (mileposts).

Dassié something of a pioneer regarding the methods of accurate identification of these itinerary distances. He extensively uses aerial photography in conjunction with global positioning satellite systems (GPS), then verifies these distances on large-scale maps, making all due allowances for curvature etc. He has proven that the Roman inscriptions relating to these itinerary distances are not all recorded in terms of the Roman league. He has accurately identified a variety of distances that were termed "leagues", because they have been repetitively found in the course of many hundreds of comparisons. He notes the earliest researches into these distances in Gaul were conducted by Bourguignon d' Anville in 1760, who calculated from the distances between the cities of Gaul a Roman league that equates to 2,211 metres. The Standard Canonical value of the Roman foot is 0.96768ft and 7,500 of them equal 2,212 metres. But the methods employed by him for his estimates of the distance are considered unreliable. However, he knew of this Roman value, it is one of those recorded by Michell. These Roman variations that concern us here are as follows:

ROOT — 176/175 — ROOT CANONICAL – 176/175 – ROOT GEOGRAPHIC
0.96 | 0.965485 | 0.971002

| 441/440 |

(All dimensions are given in English feet)

STANDARD — 176/175 — STD CANONICAL — 176/175 — STD GEOGRAPHIC
0.9621818 | 0.96768 | 0.9732096

- THE METROLOGY OF THE ROMAN FOOT -

Using more methodical techniques, in 1770, De La Sauvagiere arrived at a value of 2,225 metres for the Roman league and 7,500 feet of the Standard Geographic classification is 2,224.75 metres. This Standard Geographic classification is often found in the itinerary modules in Britain.

Pistollet de Saint-Ferjeux, in 1858 becomes the first to propose a longer league of pre Roman origin. He is stated to have calculated the league as 2,415 metres, and one and a half English miles is 2414 metres. Therefore the basic foot may be stated to have been 1.056ft - which is the Root Canonical value of the Persian foot and directly related to the original *pied de roi* by the familar ratio of 1:1.008. In 1865, Auries, the author of 14 memoirs concerning the Gallic league, proposed a value of 2,436 metres.

Lievre later confirmed this distance in 1893; and described the methods he used in this thorough determination following the route from Tours to Poitiers, a distance of 102.3 kilometres. He consulted the *Tables of Peutinger*, which stated that this distance was 42 leagues giving the same length for the league as Lievre's 2,436 metres. This lies within $2\frac{1}{2}$ metres of the original reckoning of the *pied de roi* adopted by Charlemagne. This may be regarded as total accuracy; the margin of error over the entire distance of nearly 64 miles is within 100 metres, the original positions of the milliary posts being debatable.

Due to the fact that this same level of accuracy is displayed by distances between all of the major cities of Gaul, as measured by Dassié from information given on the Roman milliary columns, it proves two major claims. Firstly the extreme longevity of measures defined in remote antiquity and surviving into modern times, secondly the accuracy of the ancient surveyors being equal to those of the modern. It further demonstrates how the extremely close values of the original feet become distinctly measurable when reaching these itinerary multiplications. For example, the $1\frac{1}{2}$ English mile league is 2,414m and the league of the Persian foot is 2,436m, an easily measured 20m longer. The two lengths divide to reveal the ratio between the Persian foot and the *pied de roi* as precisely 1 : 1.008. This is not unexpected; one is root canonical, the other standard geographical value.

THE BELGIC FOOT

In 15 BC, Nero Claudius Drusus, brother of Tiberius, became governor of the states north of Italy. Gaul had three distinct divisions; they encompassed all of France and overlapped what are now parts of Spain, Germany and the Low Countries. He adopted the foot of the Germanic Tungri as an exchange linear standard with these northern territories and Rome. Their administrative centre was at Tongeren, in what is now Belgium, where their standards were kept. The reason for this adoption is that the Belgic foot was comprised of 18 digits to the 16 digits of the Roman foot. This was a unit already familiar to the Romans through the similarity of Greek system in which the 18 digits module was termed a pygme. This foot had been widely used in England, indeed, southern England had been inhabited by these Belgic tribes one of whom were the Atribates who had supported their relatives in the Gallic wars.

Heath and Michell refer to this foot as the Drusian foot, it is basic to their interpretations of the elaborate ancient patterns drawn upon the English landscape and an examination of its pedigree will greatly substantiate their claims. Its values relevant to this examination are given in the table below, these being exactly 9 to 8 of the Roman.

ROOT	176/175	ROOT CANONICAL	176/175	ROOT GEOGRAPHIC
1.08		1.08617143 (The 'Drusian Foot')		1.092378
441/440				All dimensions are given in English feet
STANDARD	176/175	STD CANONICAL	176/175	STD GEOGRAPHIC
1.082454		1.08864		1.0948608

- THE METROLOGY OF THE BELGIC FOOT -

This so called Belgic foot had a wide area of dispersal. Oppert had recorded a foot of 1.08ft during his excavations at Babylon, dating from the Assyrian period. Josephus, in describing the 'amah' of the Temple of Jerusalem stated the ratio of the amah to the Roman cubit is 3/2. Thus the amah may be regarded as a two feet cubit of the Belgic foot, and at the above Root Canonical classification, is the exact same cubit that is prominent in the Temple of Stonehenge and the prehistoric survey line to Lundy, according to the findings of Michell and Heath.

We know from the records of Nero Claudius Drusus that the exchange for the Belgic foot was taken to be nine eighths of the Roman. Therefore, 8/9 of the Belgic foot at Standard Geographic is 0.9732096ft , which is the Standard Geographic Roman foot, precisely one of the values of the Roman foot identified by Petrie himself, and which he termed a Pelasgo foot. It is this foot, which is the basis of the Roman mile of which 75 equal a mean geographic degree. It is the foot of which one hundred comprise the inner diameter of the sarsen lintel circle at Stonehenge.

Once the system of classification in ancient metrology are understood, it makes available a new tool with which to clarify many areas of historical research. The lineage of all ancient measure is ratio and fraction related, as the above examples amply demonstrate.

THE BRITISH MEASURES PROPOSED HEREIN

The long distance measurements proposed by Heath and Michell exactly conform to ancient units familiar to metrologists and confirmed elsewhere. Their findings cannot be censured as contrived or romantic, as one could perhaps criticise such proposed landscape effigies as the Glastonbury Zodiac. Their discoveries are more in the nature of something one would expect to find as the result of a detailed and practical survey. After millennia, enough of the ancient trig. points still remain to be identified.

The itinerary modules found between Stonehenge and Silbury Hill correspond with previously found examples to the almost unbelievable levels of accuracy that one comes to associate with the Bronze Age science exposed here. Michell gives it as 86,893 feet, and in terms of accepted historical itineraries, this may be expressed as either 18 Roman miles or 12 Roman leagues exactly. The value of the Roman foot in this equation is the Root Canonical 0.96548ft. Alternative interpretations exist in terms of other ancient measures. For example, it is ten 7,500ft leagues in terms of the Royal Egyptian foot.

This view, that of the intended distance between Stonehenge and Silbury, being in terms of 18 Roman miles or 12 Roman leagues, is further strengthened by the fact that the latitudinal distance between the sites of Stonehenge and Avebury, is exactly 18 Greek miles and 12 Greek leagues in terms of the correct geographic foot, 1.01376ft, that fit this degree of latitude. These observations alone give credibility to the proposal made earlier that the Romans had little or nothing to do with the long distance alignments found all over Britain. It becomes increasingly obvious that the measurement system adopted by them was an inheritance of great antiquity, from a source identical to that of the inhabitants of Britain.

The length of Heath's proposed extension of the Stonehenge station rectangle to Lundy Island is also compatible with these Roman and Greek miles and leagues that connect Silbury with the Henge. The scale of the Stonehenge rectangle to its extension is 1:2500; this is an overall distance of 651,702ft. This length is then exactly 135 Root Canonical Roman miles or 90 leagues; it would also yield a canonical number solution in terms of the Greek measure, being 129.6 miles and 86.4 leagues. Furthermore the distance is also 75 royal Egyptian miles or 50 leagues, in terms of the royal Egyptian foot.

Because we are using such a fractionally integrated series of modules over these distances, then many alternative interpretations become possible.However, I do not believe any of the above interpretations to be the design intention here. Given that the original Stonehenge rectangle is compatible in its dimensions, and above all, its ratios, with the Belgic foot of 1.086171ft, then it is obvious to seek the solution of the intended overall design in terms of this foot.

The ratio of the Stonehenge rectangle being five, twelve, thirteen becomes 50, 120 and 130 in terms of units of two of these feet. If the sides of greater rectangle are divided by these ratios then, at this scale, two Belgic feet at the Henge is exactly one 5000ft Belgic mile in the extension triangle. Consequently the distance from Stonehenge to Lundy is 120 Belgic miles, from Lundy to the Preseli Mountains is 50 Belgic miles, and the distance to Stonehenge from Preseli is 130 of these miles.

Not only is this a perfect design, but it is also a perfectly executed design.

THE PERPETUAL CHOIRS DECAGON

This rationality continues, by now it seems remorselessly, through to the final pattern discussed in the book, the great decagon of the perpetual choirs. Centred upon White-leaved Oak, the meeting point of the three counties of Gloucester, Worcester and Hereford, the radius of this circle is a little less than 62.5 statute miles. In terms of the mile of 5,000 'Sumerian' feet of the Standard classification, it is exactly 60 miles or 40 leagues. (There would seem to a precedent for this mile from the *Antonine Itinerary*, given as the 17 miles between High Cross and Whilton Lodge, the distance is very nicely 17 x 5,000 'Sumerian' feet).

One tenth of the perimeter of the perpetual choirs circle, in terms of a mile, would be 40 miles of the common Greek foot 1.0368ft. However, the straight lines forming the decagon are the important distances in this figure, and at 204,120ft are extraordinarily interesting, in as much as this identical length has been independently identified by Dassié as one of the Gallic leagues. At 2,488.63 metres to the league, this distance is exactly 25 leagues of 7,500ft, the constituent foot being in this case the Belgic Standard Canonical 1.09964 feet.

Dassié quotes variable, but precise, values for the league of between 2,400 and 2,500 metres. At the lower end of the scale, these leagues are expressed in terms of the root 1.05 ft 'Persian' measure, and the upper, the root 1.08 ft 'Belgic' measure.

SUMMARY

THE ADVANTAGES OF USING ANCIENT METROLOGY IN ARCHAEOLOGICAL INVESTIGATION

Applying the Roman information from maps and milestones to absolute values of the given leagues has enabled Dassié to locate lost Roman settlements. In a single province (Charente Maritime) he has discovered: *Tamnum,* in Consac, *Lamnun,* in Pons, and *Novioregum,* in Barzan. This is an invaluable spin-off from the study of metrology, which could be advantageously and directly applied by archaeologists elsewhere.

The techniques employed by Heath and Michell in their research into ancient surveying in Britain have a pedigree based on an understanding of ancient metrology. In *The Measure of Albion* the reader may enjoy the first-fruits of the scientific integration of metrology into prehistoric research in Britain.

As well as metrology being an awe-inspiring pursuit in itself, Heath and Michell are using it as a most useful practical tool in ground-breaking research. I hope that this appendix will convince at least some readers that it is high time that the subject was resurrected from its present obscurity.

APPENDIX TWO - A GUIDE TO PREHISTORIC GEODETICS

Field-work Calculations, Problems and Solutions

When John asked me for my estimate of the distance between Stonehenge and Lundy during our meeting in London, I had complete confidence in the figure I gave to him, 123.4 miles. This distance was based on the coordinates of Stonehenge and those of Lundy centre, taken from OS Maps, and calculated using the formulae found in the leading *modern* works on geodesy, adapted to incorporate the *ancient* values for the size of the earth. At this point I obtained 123.427 miles *(see example one below)*. This suggested, for all the reasons espoused in this book, that accurate surveying had taken place in prehistory.

EXAMPLE ONE

FINDING THE DISTANCE FROM STONEHENGE TO LUNDY CENTRE

Latitude of both sites: 51° 10′ 42″N   Cos(Latitude) = 0.626898476

Equatorial circumference 24,902.94857 miles

Small circle circ. (lat 51° 10′ 42″N) = 24,902.94857142 x cos(lat) = 15,611.62053 miles.

Difference in longitude between sites: = 4° 40′ 15.3″ - 1° 49′ 29″ = 2.846194444°

Length of one degree of long. at lat. 51° 10′42″ = 15,611.62053/360 = 43.365561258 miles.

Distance = 43.36556126 x 2.846194444° = 123.42697 miles.

This lies within 8.5 feet of 123.42857 miles

This calculation assumes a spherical form for the earth, while the earth's shape is an ellipsoid. It appears that the ancient surveyors applied the calculation above, as the ellipsoid model would alter distances by a tiny yet significant amount. The formula for calculating the small circle circumference on an ellipsoidal earth is given on page 137.

USING MAPS

On any map of southern Britain, a lunation triangle drawn between Stonehenge, Lundy and the Preseli site measures 5:12:13. Allowance must be made for the slewing of north-south caused by mapping distortion, which makes the right-angle appear less than 90°, but the triangle emerges with sufficient accuracy to define the dimensions.

Unfortunately, it is not possible to measure distances from an OS map accurately enough for our purposes using a ruler. The map projection used by the OS contains linear distortion - a straight line other than true north-south drawn across the map represents a shallow curve on the territory. It may also be tempting to assume that ancient sites marked on an OS map are positioned precisely, but this is sometimes not the case, making it essential to visit the sites in question. Then, the careful use of an OS map using the techniques exampled here will enable a researcher to establish the latitude and longitude of a site within a second (100 feet), certainly accurately enough to discover if the distances between ancient sites spaced by more than a few miles were struck with a known historic measure (*see appendix one for ancient capabilities here.*)

EXAMPLE TWO

In this example, the two points involved are not on the same latitude or longitude, point a is south-east of point b. The calculation involves an extra complication, *stage three,* which the following worked example should clarify.

FINDING THE DISTANCE BETWEEN STONEHENGE AND GLASTONBURY

Referring to the picture on page 104 we discover that Avebury is to the west of Stonehenge by a very small angle from north. The coordinates of the centre of Stonehenge are latitude 51° 10′ 42″N; longitude 1° 49′ 29.11″ W, those of Arthur's Grave, Glastonbury are as given below.

Stage One : convert coordinates to DD (decimal degree format).

Stonehenge latitude 51.17840156° N ; longitude 1.824813448° W
Glastonbury latitude 51.14638889° N ; longitude 2.7150° W

Difference in latitude = 51.42869444 - 51.17840156 = 0.03201267°
Mean latitude = (lat. Stonehenge + lat. Cove) /2 = 51.16239523°
Difference in longitude = 2.7150 - 1.824813448 = 0.890186552°

Stage Two : Convert these angular distances to lengths.

Latitude difference = length per degree of lat. x angular distance = 69.12 miles x 0.03201267°
= 2.21271575 miles.
Longitude difference = length per degree of long. x angular distance = 24,902.94857143 miles/360 x cos(mean latitude) x 0.890186552° = 38.61682921 miles.

Stage Three : Solve the right angled triangle to find the required distance.

Because we now have two sides of a right angled triangle, the east-west distance (A) and the north-south distance (B), we can now apply Pythagoras's theorem to find the length of the hypotenuse, the sought after distance between the two points (C).

$$A^2 + B^2 = C^2$$

1491.259498 + 4.89611099 = 1496.155609

C= (square root of 1496.155609) = 38.6802 miles *Answer*.

The bearing is given by the angle whose tangent is A/B.

$\tan^{-1} A/B$ = 3.279° south of west *Answer* (azimuth = 266.721°).

In this calculation the assumption is made that the points lie on a flat surface. For distances less than about fifty miles, this assumption leads to a minute error, the effect of the earth's curvature being small. However, these errors are not negligeable for larger distances, nor unimportant for the kind of research undertaken here. The reader will need to understand spherical trigonometry to estimate the difference between results, and the formula for spherical triangle work, plus a worked example showing how it is applied, are given overleaf.

The following ellipsoid and spherical formulae may be of use:

The circumference of a small circle at a given latitude on an ellipsoid is equal to,

$$\frac{2\pi \cos(lat)\, a(1-c)}{[\cos^2(lat) \times (1-c)^2 + \sin^2(lat)]^{-2}}$$ [where a = equatorial radius, b = polar radius, c = (a - b)/a]

The approx length of the meridian degree, for any given latitude, is given by the formula,

$$\frac{[6077 - 31 \cos\{2(latitude)\}]}{88} \text{ miles}$$

General formula for a spherical triangle:

The three angles of the triangle are A, B and C

The three sides of the triangle are a,b and c.

(These are expressed as angles of arcs of great circles with respect to the centre of the earth. One of the points of the triangle is taken to be the pole).

Then cos b = cos a.cos c + sin a. sin c. cos B

EXAMPLE 3 : CALCULATE THE DISTANCE BETWEEN LONDON AND NEW YORK

London (51° 30′ N; 0° W); New York (41° N; 74° W).

Using the pole (B) as the third point of the triangle, find the co-latitude of London (A) and New York (C).

a = $90^{\circ} - 51.5^{\circ} = 38.5^{\circ}$; c = $90^{\circ} - 41^{\circ} = 49^{\circ}$; B (difference in longitude) = $74^{\circ} - 0^{\circ} = 74^{\circ}$

Then cos c = 0.656059029 ; cos a = 0.782608156 ; sin c = 0.75470958 ; sin a = 0.622514636 and cos B = 0.275637355.

Thus, from the formula above,

cos b = (0.656059029 x 0.782608156) + (0.75470958 x 0.622514636 x 0.275637355)
= 0.513437147 + 0.129499324
= 0.642936471.

whence b = 49.98886506 , which multiplied by the average meridian degree 69.12, yields the answer as 3,455.230 miles.

ESTIMATING LATITUDE AND LONGITUDE

The most accurate way to establish distance between two points is to determine their latitude and longitude, either read from the margins of the map (some modern OS maps omit this essential aid) or taken on site from a global positioning (GPS) device set to the OS geoid. Then apply the calculations given here.

The absolute determination of latitude became possible, using sophisticated instrumentation and techniques perfected during the nineteenth century, primarily by Col. Talcott in the USA . The accurate determination of longitude came about following the development of better chronometers. Harrison's chronometer set the pace for such devices, and the modern radio chronometer or caesium clock provides ultimate accuracy.

The GPS device is a new development, and gives a direct reading of the latitude and longitude of a site to within 20 or 30 feet. Because of this 'within 20 feet' spread, the measurement of smaller distances between sites spaced less than a few thousand feet apart cannot be accurately calculated by GPS, which introduces progressively more and more error and is thereby rendered useless. For such measurements, a good fibreglass tape measure is still essential kit, together with a high quality compass. In addition, the authors have found slight but significant discrepancies between the displayed coordinates of the GPS device and those marked on an OS map. The GPS device is a useful tool in this work, carefully applied.

ESTIMATING ANGLES

For absolute determination of angles and azimuths (bearings), it is best to take readings from a good theodolite, but it is recognised that this is beyond most casual researchers. Remote sites which precluded access with vehicles and/or involve hill climbing convinced the author many years ago that this instrument would have been better named the theodoheavy! One viable alternative is to use the GPS to establish a north-south line over a few hundred yards, marking its ends with posts. A bearing accurate to within a degree is easily possible using a 6″ protractor, by referring to this line. Some GPS devices calculate the bearing between two points more accurately than this, if not the calculation can always be done on site with a pocket calculator *(see example two)*. What is certain is that the researcher will end up walking miles, often across difficult territory.

USING COMPUTER PROGRAMS TO CALCULATE GEODETIC PROBLEMS

The pocket calculator rapidly becomes tiresome for repeated calculations of those exampled above. There exist programs for computers which offer distances and angles between points quickly and easily, but here we must issue a warning. Many of these programs are accurate approximations, inadequate for working with the ancient system unless one can change the constants used to generate the answer. The author has written a range of programs which may be available from from a website in the near future.

APPENDIX THREE - LOCATIONS OF SITES MENTIONED IN THE BOOK

AVEBURY :
The Sanctuary (centre) 51° 24′ 38″N ; 1° 49′ 47″W
West Kennett Longbarrow (entrance) 51° 24′ 30″N ; 1° 50′ 56″W
The Cove 51° 25′ 43″N ; 1° 51′ 09″W
Obelisk 51° 25′ 39″N ; 1° 51′ 06″W
Silbury Top 51° 24′ 56″ N ; 1° 51′ 24″W
Alton Priors church 51° 21′ 27″N ; 1° 50′ 36″W

STONEHENGE
Centre 51° 10′ 42.4″N ; 1° 49′ 29.1″W
Heelstone 51° 10′ 46″N ; 1° 49′ 26″W
Airman's monument 51° 11′ 06″N ; 1° 51′ 33″W

Gare Hill 51° 09′ 37″N ; 2° 18′ 56″W

GLASTONBURY
St Mary Chapel pillar 51° 08′ 46.5″N ; 2° 42′ 53″W
St Michael's tower, Tor 51° 08′ 39.5″N ; 2° 41′ 50″W
Barrow Mump 51° 04′ 12″N ; 2° 54′ 53″W

PRESELI Carn Wen Summit (trig point) 51° 55′ 26″N ; 4° 39′ 56″W
LUNDY centre tump 51° 10′ 42″N ; 4° 40′ 15″W
CALDEY Priory church 51° 38′ 03″N ; 4° 41′ 12″W

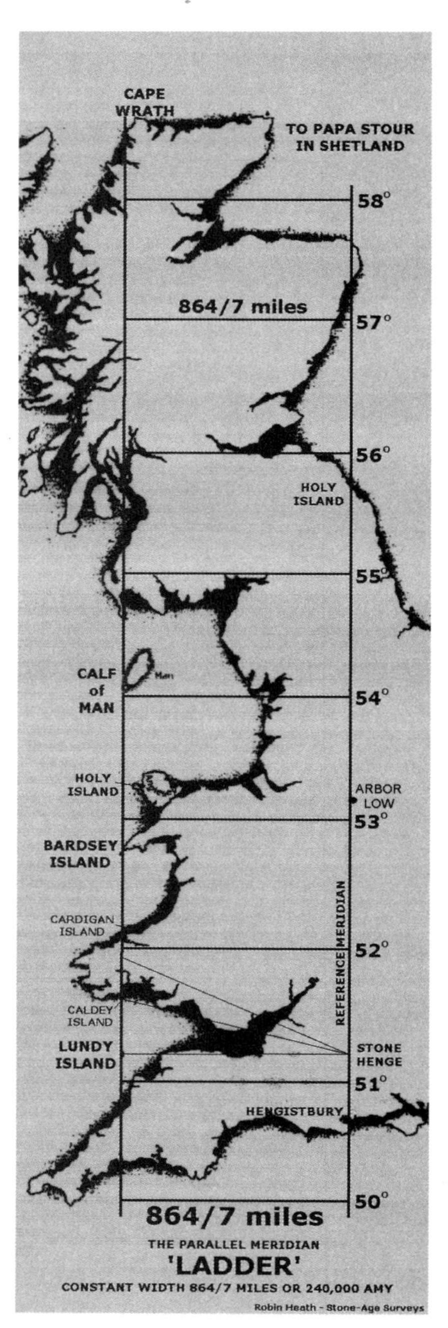

APPENDIX FOUR - THE PARALLEL MERIDIAN 'LADDER'

The north-south meridian at Stonehenge is taken as the reference meridian, the meridian to its left projected as a parallel, at a constant distance of 864/7 miles. This left hand leg of the ladder splays out to the left as it travels northwards, and is not a true meridian.

This projection technique, sometimes called *Plane Chart mapping,* appears to define a major component of the prehistoric survey revealed in this book. It forms the axis line for Caesar's Triangle, while the distance from Stonehenge to Lundy Island defines the width of the 'rungs' as a constant 864/7 miles, or 240,000 AMYs.

The left-hand leg also runs up through the chain of islands conveniently placed along the west coast of Britain - Lundy, Caldey, Cardigan Island (off axis), Bardsey, Holy Island, Anglesey (off axis), Calf of Man and on up to Cape Wrath. Stonehenge continues this theme, the central meridian passing very close to Holy Island (near Lindesfarne), skirting the east Aberdeenshire coast and striking up past Papa Stour in the West Shetlands.

The locations of the individual Islands and other key points along the ladder are given below. Note that Arbor Low, the 'Stonehenge of the North' and dated at 3000 BC, is almost exactly 2° north of Stonehenge (53° 10′ N; 1° 46′W).

COORDINATES

Stonehenge 51° 10′ 42″N ; 1° 49′ 29″W
Lundy Centre 51° 10′ 42″N ; 4° 40′ 15″W
Caldey (Tumulus)51° 38′ 00″N ; 4° 41′ 58″W
Cardigan Island 52° 08′ 00″N ; 4° 41′ 20″W
Bardsey Island 52° 45′ 40″N ;4° 47′ 23″W
Holy Is., Anglesey 53° 19′ 45″N ; 4° 41′ 35″W
Calf of Man 54° 03′ 12″N ; 4° 48′ 30″W
Cape Wrath 58° 37′ 30″ N ; 4° 59′ 54″ W
Holy Is.,(Lindesfarne) 55° 41′ N ; 1° 47′ W
Peterhead 57° 32′ N ; 1° 47′ W
Papa Stour(Shetland) 60° 19′ N ; 1° 40′W

APPENDIX FIVE - BIBLIOGRAPHY
Books referred to or relevant to the text

BOOKS BY JOHN MICHELL

Dimensions of Paradise (Thames & Hudson)
A New View over Atlantis (Thames & Hudson)
At the Centre of the World (Thames & Hudson)
Twelve Tribe Nations (with Christine Rhone) (Thames & Hudson)
A Little History of Astroarchaeology (Thames & Hudson)
Megalithomania (Thames & Hudson)
Ancient Metrology (Pentacle Books)

BOOKS BY ROBIN HEATH

A Key to Stonehenge (Bluestone Press)
Stone Circles : A Beginner's Guide (Hodder Headline)
Sun, Moon & Earth (Wooden Books, also Walker & Co, New York)
Stonehenge (Wooden Books, also Walker & Co, New York)
Sun, Moon & Stonehenge (Bluestone Press)

BOOKS ON SURVEYING/GEODESY/ASTRONOMY

Plane and Geodetic Surveying, Vol I & II, Prof. David Clark (Glasgow University)
The Astronomical and Mathematical Foundations of Geography, Charles H Cotter (Hollis & Carter).

BOOKS ON MEGALITHIC SCIENCE

Megalithic Sites in Britain, Prof. Alexander Thom (Oxford)
Stonehenge, Prof. John North (Harper Collins)
The Keys to the Temple, David Furlong (Piatkus)
Uriel's Machine, Dr Robert Lomas and Dr Christopher Knight (Century)
Science & Society in Prehistoric Britain, Dr Euan MacKie (Elek)
Beyond Civilisation, Dr Colin Renfrew (Cambridge)
The Matrix of Creation, Richard Heath (Bluestone Press)
Stonehenge Decoded, Dr Gerald Hawkins (Souvenir Press)

BOOKS ON PREHISTORIC ANTHROPOLOGY

Blood Relations, Dr Chris Knight (Yale)

BOOKS ON METROLOGY

Historical Metrology, Prof Algernon Berriman (Dent)
All Done with Mirrors, J F Neal (Secret Academy)

BOOKS ON EARTH MYSTERIES

Earth Magic, Francis Hitching (Book Club Assoc)
The Sun and the Serpent, Paul Broadhurst & Hamish Miller (Pendragon Press)
The Dance of the Dragon, Paul Broadhurst & Hamish Miller (Pendragon Press)
Earthlights, Paul Devereux (Turnstone Press)
The Old Straight Track, Alfred Watkins (reprinted by Garnstone Press)

APPENDIX SIX:
THE COORDINATES OF THE PERPETUAL CHOIRS DECAGON

The coordinates are calculated using a base line defined by one side of the decagon, that between Glastonbury (Mary Chapel) and Stonehenge. A 'closed traverse' calculation, as per example two in appendix two, returned to Glastonbury after a circuit around the decagon, and having calculated all the other eight coordinates, produced a residual distance error of less than 380 feet. The towns listed below (*) are those nearest to the actual locations. The coordinates of the central point, Whiteleafed Oak, are 52° 01′ 20″N ; 2° 21′ 03″W.

1. *Glastonbury*	51° 08′ 47″N ; 2° 42′ 54″W
2. *Stonehenge*	51° 10′ 42″N ; 1° 49′ 29″W
3. *Goring*	51° 31′ 56″N ; 1° 07′ 58″W
4. *Stony Stratford**	52° 04′ 01″N ; 0° 54′ 08″W
5. *Croft Hill*	52° 35′ 17″N ; 1° 13′ 55″W
6. *Uttoxeter**	52° 53′ 25″N ; 1° 59′ 45″W
7. *Ellesmere**	52° 51′ 19″N ; 2° 54′ 08″W
8. *Carno /Gors Goch**	52° 30′ 15″N ; 3° 36′ 16″W
9. *Llandovery*	51° 57′ 47″N ; 3° 50′ 06″W
10 *Llantwit major*	51° 26′ 31″N ; 3° 30′ 20″W

INDEX

Index

Index

Index

Opposite: The authors surveying from the Ridgeway, exactly north of Stonehenge, at latitude $^{360}/_{7}\,^{\circ}$, near Avebury, November 2002.

(photograph by Tricia Osborne)

NASA, NAZIS & JFK:
The Torbitt Document & the JFK Assassination
introduction by Kenn Thomas

This book emphasizes the links between "Operation Paper Clip" Nazi scientists working for NASA, the assassination of JFK, and the secret Nevada air base Area 51. The Torbitt Document also talks about the roles played in the assassination by Division Five of the FBI, the Defense Industrial Security Command (DISC), the Las Vegas mob, and the shadow corporate entities Permindex and Centro-Mondiale Commerciale. The Torbitt Document claims that the same players planned the 1962 assassination attempt on Charles de Gaul, who ultimately pulled out of NATO because he traced the "Assassination Cabal" to Permindex in Switzerland and to NATO headquarters in Brussels. The Torbitt Document paints a dark picture of NASA, the military industrial complex, and the connections to Mercury, Nevada which headquarters the "secret space program."

258 PAGES. 5X8. PAPERBACK. ILLUSTRATED. $16.00. CODE: NNJ

INSIDE THE GEMSTONE FILE
Howard Hughes, Onassis & JFK
by Kenn Thomas & David Hatcher Childress

Steamshovel Press editor Thomas takes on the Gemstone File in this run-up and run-down of the most famous underground document ever circulated. Photocopied and distributed for over 20 years, the Gemstone File is the story of Bruce Roberts, the inventor of the synthetic ruby widely used in laser technology today, and his relationship with the Howard Hughes Company and ultimately with Aristotle Onassis, the Mafia, and the CIA. Hughes kidnapped and held a drugged-up prisoner for 10 years; Onassis and his role in the Kennedy Assassination; how the Mafia ran corporate America in the 1960s; the death of Onassis' son in the crash of a small private plane in Greece; Onassis as Ian Fleming's archvillain Ernst Stavro Blofeld; more.

320 PAGES. 6X9 PAPERBACK. ILLUSTRATED. $16.00. CODE: IGF

POPULAR PARANOIA
The Best of Steamshovel Press
edited by Kenn Thomas

The anthology exposes the biologocal warfare origins of AIDS; the Nazi/Nation of Islam link; the cult of Elizabeth Clare Prophet; the Oklahoma City bombing writings of the late Jim Keith, as well as an article on Keith's own strange death; the conspiratorial mind of John Judge; Marion Pettie and the shadowy Finders group in Washington, DC; demonic iconography; the death of Princess Diana, its connection to the Octopus and the Saudi aerospace contracts; spies among the Rajneeshis; scholarship on the historic Illuminati; and many other parapolitical topics. The book also includes the Steamshovel's last-ever interviews with the great Beat writers Allen Ginsberg and William S. Burroughs, and neuronaut Timothy Leary, and new views of the master Beat, Neal Cassady and Jack Kerouac's science fiction.

308 PAGES. 8X10 PAPERBACK. ILLUSTRATED. $19.95. CODE: POPA

MIND CONTROL, OSWALD & JFK:
Were We Controlled?
introduction by Kenn Thomas

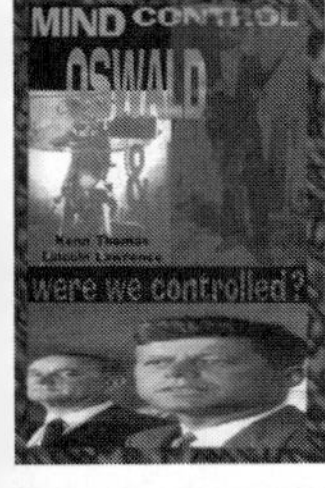

Steamshovel Press editor Kenn Thomas examines the little-known book *Were We Controlled?*, first published in 1968. The book's author, the mysterious Lincoln Lawrence, maintained that Lee Harvey Oswald was a special agent who was a mind control subject, having received an implant in 1960 at a Russian hospital. Thomas examines the evidence for implant technology and the role it could have played in the Kennedy Assassination. Thomas also looks at the mind control aspects of the RFK assassination and details the history of implant technology. Looks at the case that the reporter Damon Runyon, Jr. was murdered because of this book.

256 PAGES. 6X9 PAPERBACK. ILLUSTRATED. NOTES. $16.00. CODE: MCOJ

THE SHADOW GOVERNMENT
9-11 and State Terror
by Len Bracken, introduction by Kenn Thomas

Bracken presents the alarming yet convincing theory that nation-states engage in or allow terror to be visited upon their citizens. It is not just liberation movements and radical groups that deploy terroristic tactics for offensive ends. States use terror defensively to directly intimidate their citizens and to indirectly attack themselves or harm their citizens under a false flag. Their motives? To provide pretexts for war or for increased police powers or both. This stratagem of indirectly using terrorism has been executed by statesmen in various ways but tends to involve the pretense of blind eyes, misdirection, and cover-ups that give statesmen plausible deniability. Lusitiania, Pearl Harbor, October Surprise, the first World Trade Center bombing, the Oklahoma City bombing and other well-known incidents suggest that terrorism is often and successfully used by states in an indirectly defensive way to take the offensive against enemies at home and abroad. Was 9-11 such an indirect defensive attack?

288 PAGES. 6X9 PAPERBACK. ILLUSTRATED. $16.00. CODE: SGOV

LIQUID CONSPIRACY
JFK, LSD, the CIA, Area 51 & UFOs
by George Piccard

Underground author George Piccard on the politics of LSD, mind control, and Kennedy's involvement with Area 51 and UFOs. Reveals JFK's LSD experiences with Mary Pinchot-Meyer. The plot thickens with an ever expanding web of CIA involvement, from underground bases with UFOs seen by JFK and Marilyn Monroe (among others) to a vaster conspiracy that affects every government agency from NASA to the Justice Department. This may have been the reason that Marilyn Monroe and actress-columnist Dorothy Kilgallen were both murdered. Focusing on the bizarre side of history, *Liquid Conspiracy* takes the reader on a psychedelic tour de force. This is your government on drugs!

264 PAGES. 6X9 PAPERBACK. ILLUSTRATED. $14.95. CODE: LIQC

THE ARCH CONSPIRATOR
Essays and Actions
by Len Bracken

Veteran conspiracy author Len Bracken's witty essays and articles lead us down the dark corridors of conspiracy, politics, murder and mayhem. In 12 chapters Bracken takes us through a maze of interwoven tales from the Russian Conspiracy (and a few "extra notes" on conspiracies) to his interview with Costa Rican novelist Joaquin Gutierrez and his Psychogeographic Map into the Third Millennium. Other chapters in the book are A General Theory of Civil War; A False Report Exposes the Dirty Truth About South African Intelligence Services; The New-Catiline Conspiracy for the Cancellation of Debt; Anti-Labor Day; 1997 with selected Aphorisms Against Work; Solar Economics; and more. Bracken's work has appeared in such pop-conspiracy publications as *Paranoia*, *Steamshovel Press* and the *Village Voice*. Len Bracken lives in Arlington, Virginia and haunts the back alleys of Washington D.C., keeping an eye on the predators who run our country. With a gun to his head, he cranks out his rants for fringe publications and is the editor of *Extraphile*, described by *New Yorker Magazine* as "fusion conspiracy theory."

256 PAGES. 6X9 PAPERBACK. ILLUSTRATED. BIBLIOGRAPHY. $14.95. CODE: ACON.

PIRATES & THE LOST TEMPLAR FLEET
The Secret Naval War Between the Templars & the Vatican
by David Hatcher Childress

Childress takes us into the fascinating world of maverick sea captains who were Knights Templar (and later Scottish Rite Free Masons) who battled the Vatican, and the Spanish and Italian ships that sailed for the Pope. The lost Templar fleet was originally based at La Rochelle in southern France, but fled to the deep fiords of Scotland upon the dissolution of the Order by King Phillip. This banned fleet of ships was later commanded by the St. Clair family of Rosslyn Chapel (birthplace of Free Masonry). St. Clair and his Templars made a voyage to Canada in the year 1298 AD, nearly 100 years before Columbus! Later, this fleet of ships and new ones to come, flew the Skull and Crossbones, the symbol of the Knights Templar. They preyed on the ships of the Vatican coming from the rich ports of the Americas and were ultimately known as the Pirates of the Caribbean. Chapters include: 10,000 Years of Seafaring; The Knights Templar & the Crusades; The Templars and the Assassins; The Lost Templar Fleet and the Jolly Roger; Maps of the Ancient Sea Kings; Pirates, Templars and the New World; Christopher Columbus—Secret Templar Pirate?; Later Day Pirates and the War with the Vatican; Pirate Utopias and the New Jerusalem; more.

320 PAGES. 6X9 PAPERBACK. ILLUSTRATED. BIBLIOGRAPHY. $16.95. CODE: PLTF

THE HISTORY OF THE KNIGHTS TEMPLARS
by Charles G. Addison, introduction by David Hatcher Childress

Chapters on the origin of the Templars, their popularity in Europe and their rivalry with the Knights of St. John, later to be known as the Knights of Malta. Detailed information on the activities of the Templars in the Holy Land, and the 1312 AD suppression of the Templars in France and other countries, which culminated in the execution of Jacques de Molay and the continuation of the Knights Templars in England and Scotland; the formation of the society of Knights Templars in London; and the rebuilding of the Temple in 1816. Plus a lengthy intro about the lost Templar fleet and its connections to the ancient North American sea routes.

395 PAGES. 6X9 PAPERBACK. ILLUSTRATED. $16.95. CODE: HKT

THE LUCID VIEW
Investigations in Occultism, Ufology & Paranoid Awareness
by Aeolus Kephas

An unorthodox analysis of conspiracy theory, ufology, extraterrestrialism and occultism. *The Lucid View* takes us on an impartial journey through secret history, including the Gnostics and Templars; Crowley and Hitler's occult alliance; the sorcery wars of Freemasonry and the Illuminati; "Alternative Three" covert space colonization; the JFK assassination; the Manson murders; Jonestown and 9/11. Also delves into UFOs and alien abductions, their relations to mind control technology and sorcery practices, with reference to inorganic beings and Kundalini energy. The book offers a balanced overview on religious, magical and paranoid beliefs pertaining to the 21st century, and their social, psychological, and spiritual implications for humanity, the leading game player in the grand mythic drama of Armageddon.

298 PAGES. 6X9 PAPERBACK. ILLUSTRATED. $16.95. CODE: LVEW

MIND CONTROL, WORLD CONTROL
by Jim Keith

Veteran author and investigator Jim Keith uncovers a surprising amount of information on the technology, experimentation and implementation of mind control. Various chapters in this shocking book are on early CIA experiments such as Project Artichoke and Project R.H.I.C.-EDOM, the methodology and technology of implants, mind control assassins and couriers, various famous Mind Control victims such as Sirhan Sirhan and Candy Jones. Also featured in this book are chapters on how mind control technology may be linked to some UFO activity and "UFO abductions."

256 PAGES. 6X9 PAPERBACK. ILLUSTRATED. FOOTNOTES. $14.95. CODE: MCWC

MASS CONTROL
Engineering Human Consciousness
by Jim Keith

Conspiracy expert Keith's final book on mind control, Project Monarch, and mass manipulation presents chilling evidence that we are indeed spinning a Matrix. Keith describes the New Man, where conception of reality is a dance of electronic images fired into his forebrain, a gossamer construction of his masters, designed so that he will not—under any circumstances—perceive the actual. His happiness is delivered to him through a tube or an electronic connection. His God lurks behind an electronic curtain; when the curtain is pulled away we find the CIA sorcerer, the media manipulatorÖ Chapters on the CIA, Tavistock, Jolly West and the Violence Center, Guerrilla Mindwar, Brice Taylor, other recent "victims," more.

256 PAGES. 6X9 PAPERBACK. ILLUSTRATED. INDEX. $16.95. CODE: MASC

PERPETUAL MOTION
The History of an Obsession
by Arthur W. J. G. Ord-Hume

Make a machine which gives out more work than the energy you put into it, and you have perpetual motion. The deceptively simple task of making a mechanism which would turn forever fascinated many an inventor, and a number of famous men and physicists applied themselves to the task. Despite the naivete and blatant trickery of many of the inventors, there are a handful of mechanisms which defy explanation. A vast canvas-covered wheel which turned by itself was erected in the Tower of London. Another wheel, equally surrounded by mystery and intrigue, turned endlessly in Germany. Chapters include: Elementary Physics and Perpetual Motion; Medieval Perpetual Motion; Self-moving Wheels and Overbalancing Weights; Lodestones, Electro-Magnetism and Steam; Capillary Attraction and Spongewheels; Cox's Perpetual Motion; Keely and his Amazing Motor; Odd Ideas about Vaporization and Liquefaction; The Astonishing Case of the Garabed Project; Ever-Ringing Bells and Radium Perpetual Motion; Perpetual Motion Inventors Barred from the US Patent Office; Rolling Ball Clocks; Perpetual Lamps; The Perpetuity of the Perpetual Motion Inventor; more.

260 PAGES. 6X9 PAPERBACK. ILLUSTRATED. BIBLIOGRAPHY. INDEX. $20.00. CODE: PPM

HIDDEN NATURE
The Startling Insights of Viktor Schauberger
by Alick Bartholomew, foreword by David Bellamy

Victor Schauberger (1885-1958) pioneered a new understanding of the Science of Nature, (re)discovering its primary laws and principles, unacknowledged by contemporary science. From studying the fast flowing streams of the unspoilt Alps, he gained insights into water as a living organism. He showed that water is like a magnetic tape; it can carry information that may either enhance or degrade the quality of organisms. Our failure to understand the need to protect the quality of water is the principle cause of environmental degradation on the planet. Schauberger warned of climatic chaos resulting from deforestation and called for work with free energy machines and energy generation. Chapters include: Schauberger's Vision; Different Kinds of Energy; Attraction & Repulsion of Opposites; Nature's Patterns & Shapes; Energy Production; Motion, Key to Balance; Atmosphere/Electricity; The Nature of Water; Hydrological Cycle; Formation of Springs; How Rivers Flow; Supplying Water; The Role of the Forests; Tree Metabolism; Soil Fertility and Cultivation; Organic Cultivation; The Energy Revolution; Harnessing Implosion Power; Viktor Schauberger & Society; more.

288 PAGES. 7X10 PAPERBACK. ILLUSTRATED. REFERENCES. INDEX. $22.95. CODE: HNAT

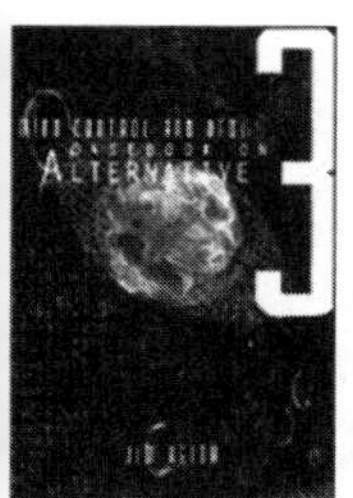

MIND CONTROL AND UFOS
Casebook on Alternative 3
by Jim Keith

Drawing on his diverse research and a wide variety of sources, Jim Keith delves into the bizarre story behind *Alternative 3,* including mind control programs, underground bases not only on the Earth but also on the Moon and Mars, the real origin of the UFO problem, the mysterious deaths of Marconi Electronics employees in Britain during the 1980s, top scientists around the world kidnapped to work at the secret government space bases, the Russian-American superpower arms race of the 50s, 60s and 70s as a massive hoax, and other startling arenas. Chapters include: Secret Societies and *Die Neuordning*; The Fourth Reich; UFOs and the Space Program; Government UFOs; Hot Jobs and Crop Circles; Missing Scientists and LGIBs; Ice Picks, Electrodes and LSD; Electronic Wars; Batch Consignments; The Depopulation Bomb; Veins and Tributaries; Lunar Base Alpha One; Disinfo; Other Alternatives; Noah's Ark II; *Das Marsprojekt*; more.

248 PAGES. 6X9 PAPERBACK. ILLUSTRATED. BIBLIOGRAPHY. $14.95. CODE: MCUF

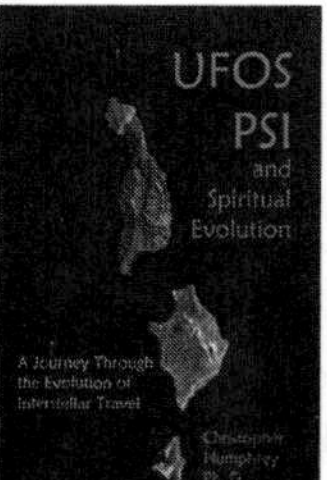

UFOS, PSI AND SPIRITUAL EVOLUTION
A Journey through the Evolution of Interstellar Travel
by Christopher Humphries, Ph.D.

The modern era of UFOs began in May, 1947, one year and eight months after Hiroshima. This is no coincidence, and suggests there are beings in the universe with the ability to jump hundreds of light years in an instant. That is teleportation, a power of the mind. If it weren't for levitation and teleportation, star travel would not be possible at all, since physics rules out star travel by technology. So if we want to go to the stars, it is the mind and spirit we must study, not technology. The mind must be a dark matter object, since it is invisible and intangible and can freely pass through solid objects. A disembodied mind can see the de Broglie vibrations (the basis of quantum mechanics) radiated by both dark and ordinary matter during near-death or out-of-body experiences. Levitation requires warping the geodesics of space-time. The latest theory in physics is String Theory, which requires six extra spatial dimensions. The mind warps those higher geodesics to produce teleportation. We are a primitive and violent species. Our universities lack any sciences of mind, spirit or civilization. If we want to go to the stars, the first thing we must do is "grow up." That is the real Journey.

274 PAGES. 6X9 PAPERBACK. ILLUSTRATED. REFERENCES. $16.95. CODE: UPSE

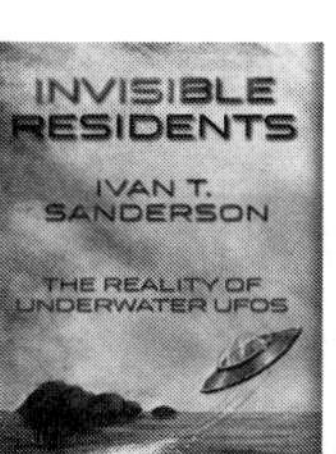

INVISIBLE RESIDENTS
The Reality of Underwater UFOS
by Ivan T. Sanderson, Foreword by David Hatcher Childress

This book is a groundbreaking contribution to the study of the UFO enigma, originally published over 30 years ago. In this book, Sanderson, a renowned zoologist with a keen interest in the paranormal, puts forward the curious theory that "OINTS"—Other Intelligences—live under the Earth's oceans. This underwater, parallel, civilization may be twice as old as Homo sapiens, he proposes, and may have "developed what we call space flight." Sanderson postulates that the OINTS are behind many UFO sightings as well as the mysterious disappearances of aircraft and ships in the Bermuda Triangle. What better place to have an impenetrable base than deep within the oceans of the planet? Yet, if UFOs, or at least some of them, are coming from beneath our oceans or lakes, does it necessarily mean that there is another civilization besides our own that is responsible? In fact, could it be that since WWII a number of underwater UFO bases have been constructed by the very human governments of our planet? Whatever their source, Sanderson offers here an exhaustive study of USOs (Unidentified Submarine Objects) observed in nearly every part of the world. He presents many well-documented and exciting case studies of these unusual sightings.; more.

298 PAGES. 6X9 PAPERBACK. ILLUSTRATED. BIBLIOGRAPHY. INDEX. $16.95. CODE: INVS

THE WORLD CATACLYSM IN 2012
Maya Calendar Countdown
by Patrick Geryl

In his previous book, *The Orion Prophecy*, author Geryl theorized that the lost civilization of Atlantis was destroyed by a huge cataclysm engendered by changes in sunspot activity affecting Earth's magnetic poles and atmosphere. Having experienced earlier catastrophes, the Atlanteans had developed amazing astronomical and mathematical knowledge that enabled them to predict the date of their continent's demise. They devised a survival plan, and were able to pass along their knowledge to civilizations we know as the Maya and Old Egyptians. Here, Geryl shows that the mathematics and astronomy of the ancient Egyptians and Maya are related, and have similar predictive power which should be taken very seriously. He cracks their hidden codes that show definitively that the next earth-consuming cataclysm will occur in 2012, and calls urgently for the excavation of the Labyrinth of ancient Egypt, a storehouse of Atlantean knowledge which is linked in prophecy to the May predictions.

256 PAGES. 6X9 PAPERBACK. ILLUSTRATED. REFERENCES. $16.95. CODE: WC20

HAARP
The Ultimate Weapon of the Conspiracy
by Jerry Smith

The HAARP project in Alaska is one of the most controversial projects ever undertaken by the U.S. Government. Jerry Smith gives us the history of the HAARP project and explains how it works, in technically correct yet easy to understand language. At best, HAARP is science out-of-control; at worst, HAARP could be the most dangerous device ever created, a futuristic technology that is everything from super-beam weapon to world-wide mind control device. Topics include Over-the-Horizon Radar and HAARP; Mind Control, ELF and HAARP; The Telsa Connection; The Russian Woodpecker; GWEN & HAARP; Earth Penetrating Tomography; Weather Modification; Secret Science of the Conspiracy; more. Includes the complete 1987 Eastlund patent for his pulsed super-weapon that he claims was stolen by the HAARP Project.

256 PAGES. 6X9 PAPERBACK. ILLUSTRATED. $14.95. CODE: HARP

SECRETS OF THE HOLY LANCE
The Spear of Destiny in History & Legend
by Jerry E. Smith and George Piccard

As Jesus Christ hung on the cross a Roman centurion pieced the Savior's side with his spear. A legend has arisen that "whosoever possesses this Holy Lance and understands the powers it serves, holds in his hand the destiny of the world for good or evil." *Secrets of the Holy Lance* traces the Spear from its possession by Constantine, Rome's first Christian Caesar, to Charlemagne's claim that with it he ruled the Holy Roman Empire by Divine Right, and on through two thousand years of kings and emperors, until it came within Hitler's grasp—and beyond! Did it rest for a while in Antarctic ice? Is it now hidden in Europe, awaiting the next person to claim its awesome power? Neither debunking nor worshiping, *Secrets of the Holy Lance* seeks to pierce the veil of myth and mystery around the Spear. Mere belief that it was infused with magic by virtue of its shedding the Savior's blood has made men kings. But what if it's more? What are "the powers it serves"?

312 PAGES. 6X9 PAPERBACK. ILLUSTRATED. BIBLIOGRAPHY. $16.95. CODE: SOHL

FROM LIGHT INTO DARKNESS
The Evolution of Religion in Ancient Egypt
by Stephen S. Mehler

Building on the esoteric information first revealed in Land of Osiris, this exciting book presents more of Abd'El Hakim's oral traditions, with radical new interpretations of how religion evolved in prehistoric and dynastic Khemit, or Egypt. * Have popular modern religions developed out of practices in ancient Egypt? * Did religion in Egypt represent only a shadow of the spiritual practices of prehistoric people? * Have the Western Mystery Schools such as the Rosicrucian Order evolved from these ancient systems? * Author Mehler explores the teachings of the King Akhenaten and the real Moses, the true identity of the Hyksos, and Akhenaten's connections to The Exodus, Judaism and the Rosicrucian Order. Here for the first time in the West, are the spiritual teachings of the ancient Khemitians, the foundation for the coming new cycle of consciousness—The Awakening; more.

240 PAGES. 6X9 PAPERBACK. ILLUSTRATED. REFERENCES. $16.95. CODE: FLID

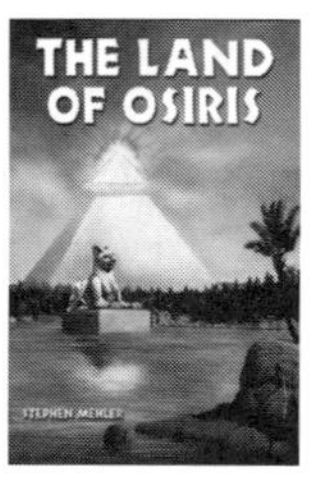

THE LAND OF OSIRIS
An Introduction to Khemitology
by Stephen S. Mehler

Was there an advanced prehistoric civilization in ancient Egypt who built the great pyramids and carved the Great Sphinx? Did the pyramids serve as energy devices and not as tombs for kings? Mehler has uncovered an indigenous oral tradition that still exists in Egypt, and has been fortunate to have studied with a living master of this tradition, Abd'El Hakim Awyan. Mehler has also been given permission to present these teachings to the Western world, teachings that unfold a whole new understanding of ancient Egypt . Chapters include: Egyptology and Its Paradigms; Asgat Nefer—The Harmony of Water; Khemit and the Myth of Atlantis; The Extraterrestrial Question; more.

272 PAGES. 6X9 PAPERBACK. ILLUSTRATED. COLOR SECTION. BIBLIOGRAPHY. $18.95. CODE: LOOS

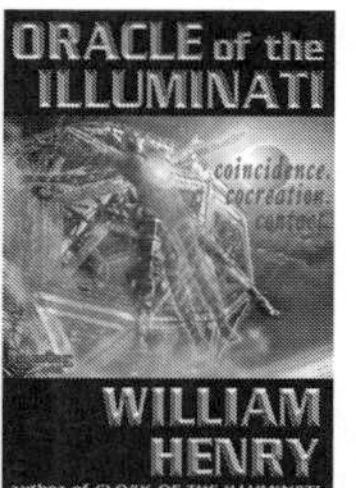

ORACLE OF THE ILLUMINATI
Coincidence, Cocreation, Contact
By William Henry

Investigative mythologist William Henry follows up his best-selling Cloak of the Illuminati with this illustration-packed treatise on the secret codes, oracles and technology of ancient Illuminati. His primary expertise and mission is finding and interpreting ancient gateway stories which feature advanced technology for raising of spiritual vibration and increasing our body's innate healing ability. Chapters include: From Cloak to Oracle; The Return of Sophia; The Cosmic G-Spot Stimulator; The Reality of the Rulers; The Hymn of the Pearl; The Realm of the Illuminati; Francis Bacon: Oracle; Abydos and the Head of Sophia; Enki and the Flower of Light; The God Head and the Dodecahedron; The Star Walker; The Big Secret; more.

243 PAGES. 6X9 PAPERBACK. ILLUSTRATED. NOTES & REFERENCES. $16.95. CODE: ORIL

CLOAK OF THE ILLUMINATI
Secrets, Transformations, Crossing the Star Gate
by William Henry

Thousands of years ago the stargate technology of the gods was lost. Mayan Prophecy says it will return by 2012, along with our alignment with the center of our galaxy. In this book: Find examples of stargates and wormholes in the ancient world; Examine myths and scripture with hidden references to a stargate cloak worn by the Illuminati, including Mari, Nimrod, Elijah, and Jesus; See rare images of gods and goddesses wearing the Cloak of the illuminati; Learn about Saddam Hussein and the secret missing library of Jesus; Uncover the secret Roman-era eugenics experiments at the Temple of Hathor in Denderah, Egypt; Explore the duplicate of the Stargate Pillar of the Gods in the Illuminists' secret garden in Nashville, TN; Discover the secrets of manna, the food of the angels; Share the lost Peace Prayer posture of Osiris, Jesus and the Illuminati; more. Chapters include: Seven Stars Under Three Stars; The Long Walk; Squaring the Circle; The Mill of the Host; The Miracle Garment; The Fig; Nimrod: The Mighty Man; Nebuchadnezzar's Gate; The New Mighty Man; more.

238 PAGES. 6X9 PAPERBACK. ILLUSTRATED. BIBLIOGRAPHY. INDEX. $16.95. CODE: COIL

THE GIZA DEATH STAR
The Paleophysics of the Great Pyramid & the Military Complex at Giza
by Joseph P. Farrell

Was the Giza complex part of a military installation over 10,000 years ago? Chapters include: An Archaeology of Mass Destruction; Thoth and Theories; The Machine Hypothesis; Pythagoras, Plato, Planck, and the Pyramid; The Weapon Hypothesis; Encoded Harmonics of the Planck Units in the Great Pyramid; High Frequency Direct Current "Impulse" Technology; The Grand Gallery and its Crystals: Gravito-acoustic Resonators; The Other Two Large Pyramids, "Causeways," and the "Temples"; A Phase Conjugate Howitzer; Evidence of the Use of Weapons of Mass Destruction in Ancient Times; more.

290 PAGES. 6X9 PAPERBACK. ILLUSTRATED. $16.95. CODE: GDS

THE GIZA DEATH STAR DEPLOYED
The Physics & Engineering of the Great Pyramid
by Joseph P. Farrell

Farrell expands on his thesis that the Great Pyramid was a chemical maser, designed as a weapon and eventually deployed—with disastrous results to the solar system. Includes: Exploding Planets: The Movie, the Mirror, and the Model; Dating the Catastrophe and the Compound; A Brief History of the Exoteric and Esoteric Investigations of the Great Pyramid; No Machines, Please!; The Stargate Conspiracy; The Scalar Weapons; Message or Machine?; A Tesla Analysis of the Putative Physics and Engineering of the Giza Death Star; Cohering the Zero Point, Vacuum Energy, Flux: Synopsis of Scalar Physics and Paleophysics; Configuring the Scalar Pulse Wave; Inferred Applications in the Great Pyramid; Quantum Numerology, Feedback Loops and Tetrahedral Physics; and more.

290 PAGES. 6X9 PAPERBACK. ILLUSTRATED. BIBLIOGRAPHY. INDEX. $16.95. CODE: GDSD

THE GIZA DEATH STAR DESTROYED
The Ancient War For Future Science
by Joseph P. Farrell

This is the third and final volume in the popular *Giza Death Star* series, physicist Farrell looks at what eventually happened to the 10,000-year-old Giza Death Star after it was deployed—it was destroyed by an internal explosion. Recapping his earlier books, Farrell moves on to events of the final days of the Giza Death Star and its awesome power. These final events, eventually leading up to the destruction of this giant machine, are dissected one by one, leading us to the eventual abandonment of the Giza Military Complex—an event that hurled civilization back into the Stone Age. Chapters include: The Mars-Earth Connection; The Lost "Root Races" and the Moral Reasons for the Flood; The Destruction of Krypton: The Electrodynamic Solar System, Exploding Planets and Ancient Wars; Turning the Stream of the Flood: the Origin of Secret Societies and Esoteric Traditions; The Quest to Recover Ancient Mega-Technology; Non-Equilibrium Paleophysics; Monatomic Paleophysics; Frequencies, Vortices and Mass Particles: the Pyramid Power of Dr. Pat Flanagan and Joe Parr; The Topology of the Aether; A Final Physical Effect: "Acoustic" Intensity of Fields; The Pyramid of Crystals; tons more.

292 PAGES. 6X9 PAPERBACK. ILLUSTRATED. BIBLIOGRAPHY. $16.95. CODE: GDES

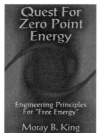

THE A.T. FACTOR

A Scientists Encounter with UFOs: Piece For A Jigsaw Part 3

by Leonard Cramp

British aerospace engineer Cramp began much of the scientific anti-gravity and UFO propulsion analysis back in 1955 with his landmark book *Space, Gravity & the Flying Saucer* (out-of-print and rare). His next books (available from Adventures Unlimited) *UFOs & Anti-Gravity: Piece for a Jig-Saw* and *The Cosmic Matrix: Piece for a Jig-Saw Part 2* began Cramp's in depth look into gravity control, free-energy, and the interlocking web of energy that pervades the universe. In this final book, Cramp brings to a close his detailed and controversial study of UFOs and Anti-Gravity.

324 PAGES. 6X9 PAPERBACK. ILLUSTRATED. BIBLIOGRAPHY. INDEX. $16.95. CODE: ATF

COSMIC MATRIX

Piece for a Jig-Saw, Part Two

by Leonard G. Cramp

Leonard G. Cramp, a British aerospace engineer, wrote his first book *Space Gravity and the Flying Saucer* in 1954. Cosmic Matrix is the long-awaited sequel to his 1966 book *UFOs & Anti-Gravity: Piece for a Jig-Saw.* Cramp has had a long history of examining UFO phenomena and has concluded that UFOs use the highest possible aeronautic science to move in the way they do. Cramp examines anti-gravity effects and theorizes that this super-science used by the craft—described in detail in the book—can lift mankind into a new level of technology, transportation and understanding of the universe. The book takes a close look at gravity control, time travel, and the interlocking web of energy between all planets in our solar system with Leonard's unique technical diagrams. A fantastic voyage into the present and future!

364 PAGES. 6X9 PAPERBACK. ILLUSTRATED. BIBLIOGRAPHY. $16.00. CODE: CMX

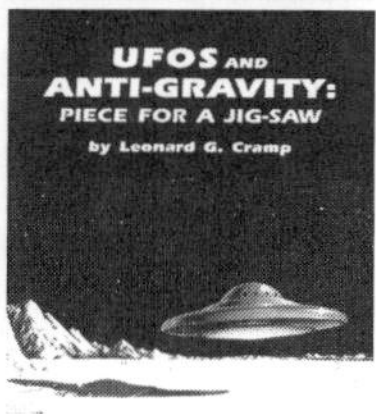

UFOS AND ANTI-GRAVITY

Piece For A Jig-Saw

by Leonard G. Cramp

Leonard G. Cramp's 1966 classic book on flying saucer propulsion and suppressed technology is a highly technical look at the UFO phenomena by a trained scientist. Cramp first introduces the idea of 'anti-gravity' and introduces us to the various theories of gravitation. He then examines the technology necessary to build a flying saucer and examines in great detail the technical aspects of such a craft. Cramp's book is a wealth of material and diagrams on flying saucers, anti-gravity, suppressed technology, G-fields and UFOs. Chapters include Crossroads of Aerodymanics, Aerodynamic Saucers, Limitations of Rocketry, Gravitation and the Ether, Gravitational Spaceships, G-Field Lift Effects, The Bi-Field Theory, VTOL and Hovercraft, Analysis of UFO photos, more.

388 PAGES. 6X9 PAPERBACK. ILLUSTRATED. $16.95. CODE: UAG

THE TESLA PAPERS

Nikola Tesla on Free Energy & Wireless Transmission of Power

by Nikola Tesla, edited by David Hatcher Childress

David Hatcher Childress takes us into the incredible world of Nikola Tesla and his amazing inventions. Tesla's rare article "The Problem of Increasing Human Energy with Special Reference to the Harnessing of the Sun's Energy" is included. This lengthy article was originally published in the June 1900 issue of *The Century Illustrated Monthly Magazine* and it was the outline for Tesla's master blueprint for the world. Tesla's fantastic vision of the future, including wireless power, anti-gravity, free energy and highly advanced solar power. Also included are some of the papers, patents and material collected on Tesla at the Colorado Springs Tesla Symposiums, including papers on: •The Secret History of Wireless Transmission •Tesla and the Magnifying Transmitter •Design and Construction of a Half-Wave Tesla Coil •Electrostatics: A Key to Free Energy •Progress in Zero-Point Energy Research •Electromagnetic Energy from Antennas to Atoms •Tesla's Particle Beam Technology •Fundamental Excitatory Modes of the Earth-Ionosphere Cavity

325 PAGES. 8X10 PAPERBACK. ILLUSTRATED. $16.95. CODE: TTP

THE FANTASTIC INVENTIONS OF NIKOLA TESLA

by Nikola Tesla with additional material by David Hatcher Childress

This book is a readable compendium of patents, diagrams, photos and explanations of the many incredible inventions of the originator of the modern era of electrification. In Tesla's own words are such topics as wireless transmission of power, death rays, and radio-controlled airships. In addition, rare material on German bases in Antarctica and South America, and a secret city built at a remote jungle site in South America by one of Tesla's students, Guglielmo Marconi. Marconi's secret group claims to have built flying saucers in the 1940s and to have gone to Mars in the early 1950s! Incredible photos of these Tesla craft are included. The Ancient Atlantean system of broadcasting energy through a grid system of obelisks and pyramids is discussed, and a fascinating concept comes out of one chapter: that Egyptian engineers had to wear protective metal head-shields while in these power plants, hence the Egyptian Pharoah's head covering as well as the Face on Mars! •His plan to transmit free electricity into the atmosphere. •How electrical devices would work using only small antennas. •Why unlimited power could be utilized anywhere on earth. •How radio and radar technology can be used as death-ray weapons in Star Wars.

342 PAGES. 6X9 PAPERBACK. ILLUSTRATED. $16.95. CODE: FINT

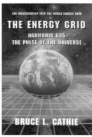

THE ANTI-GRAVITY HANDBOOK

edited by David Hatcher Childress, with Nikola Tesla, T.B. Paulicki, Bruce Cathie, Albert Einstein and others

The new expanded compilation of material on Anti-Gravity, Free Energy, Flying Saucer Propulsion, UFOs, Suppressed Technology, NASA Cover-ups and more. Highly illustrated with patents, technical illustrations and photos. This revised and expanded edition has more material, including photos of Area 51, Nevada, the government's secret testing facility. This classic on weird science is back in a 90s format!

- **How to build a flying saucer.**
- **Arthur C. Clarke on Anti-Gravity.**
- **Crystals and their role in levitation.**
- **Secret government research and development.**
- **Nikola Tesla on how anti-gravity airships could draw power from the atmosphere.**
- **Bruce Cathie's Anti-Gravity Equation.**
- **NASA, the Moon and Anti-Gravity.**

230 PAGES. 7X10 PAPERBACK. ILLUSTRATED. $14.95. CODE: AGH

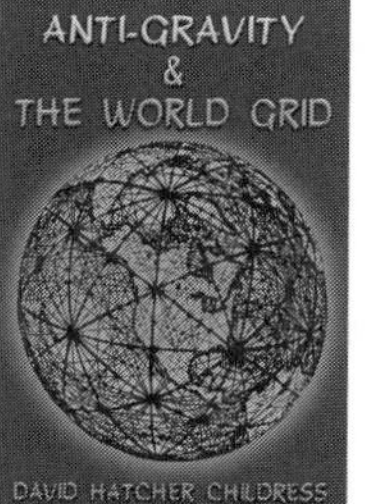

ANTI–GRAVITY & THE WORLD GRID

Is the earth surrounded by an intricate electromagnetic grid network offering free energy? This compilation of material on ley lines and world power points contains chapters on the geography, mathematics, and light harmonics of the earth grid. Learn the purpose of ley lines and ancient megalithic structures located on the grid. Discover how the grid made the Philadelphia Experiment possible. Explore the Coral Castle and many other mysteries, including acoustic levitation, Tesla Shields and scalar wave weaponry. Browse through the section on anti-gravity patents, and research resources.

274 PAGES. 7X10 PAPERBACK. ILLUSTRATED. $14.95. CODE: AGW

ANTI–GRAVITY & THE UNIFIED FIELD

edited by David Hatcher Childress

Is Einstein's Unified Field Theory the answer to all of our energy problems? Explored in this compilation of material is how gravity, electricity and magnetism manifest from a unified field around us. Why artificial gravity is possible; secrets of UFO propulsion; free energy; Nikola Tesla and anti-gravity airships of the 20s and 30s; flying saucers as superconducting whirls of plasma; anti-mass generators; vortex propulsion; suppressed technology; government cover-ups; gravitational pulse drive; spacecraft & more.

240 PAGES. 7X10 PAPERBACK. ILLUSTRATED. $14.95. CODE: AGU

THE GIZA DEATH STAR

The Paleophysics of the Great Pyramid & the Military Complex at Giza

by Joseph P. Farrell

Physicist Joseph Farrell's amazing book on the secrets of Great Pyramid of Giza. *The Giza Death Star* starts where British engineer Christopher Dunn leaves off in his 1998 book, *The Giza Power Plant*. Was the Giza complex part of a military installation over 10,000 years ago? Chapters include: An Archaeology of Mass Destruction, Thoth and Theories; The Machine Hypothesis; Pythagoras, Plato, Planck, and the Pyramid; The Weapon Hypothesis; Encoded Harmonics of the Planck Units in the Great Pyramid; High Fregquency Direct Current "Impulse" Technology; The Grand Gallery and its Crystals: Gravito-acoustic Resonators; The Other Two Large Pyramids; the "Causeways," and the "Temples"; A Phase Conjugate Howitzer; Evidence of the Use of Weapons of Mass Destruction in Ancient Times; more.

290 PAGES. 6X9 PAPERBACK. ILLUSTRATED. $16.95. CODE: GDS

DARK MOON

Apollo and the Whistleblowers

by Mary Bennett and David Percy

- Was Neil Armstrong really the first man on the Moon?
- Did you know a second craft was going to the Moon at the same time as Apollo 11?
- Do you know that potentially lethal radiation is prevalent throughout deep space?
- Do you know there are serious discrepancies in the account of the Apollo 13 'accident'?
- Did you know that 'live' color TV from the Moon was not actually live at all?
- Did you know that the Lunar Surface Camera had no viewfinder?
- Do you know that lighting was used in the Apollo photographs—yet no lighting equipment was taken to the Moon?

All these questions, and more, are discussed in great detail by British researchers Bennett and Percy in *Dark Moon*, the definitive book (nearly 600 pages) on the possible faking of the Apollo Moon missions. Bennett and Percy delve into every possible aspect of this beguiling theory, one that rocks the very foundation of our beliefs concerning NASA and the space program. Tons of NASA photos analyzed for possible deceptions.

568 PAGES. 6X9 PAPERBACK. ILLUSTRATED. BIBLIOGRAPHY. INDEX. $25.00. CODE: DMO

TECHNOLOGY OF THE GODS

The Incredible Sciences of the Ancients

by David Hatcher Childress

Popular *Lost Cities* author David Hatcher Childress takes us into the amazing world of ancient technology, from computers in antiquity to the "flying machines of the gods." Childress looks at the technology that was allegedly used in Atlantis and the theory that the Great Pyramid of Egypt was originally a gigantic power station. He examines tales of ancient flight and the technology that it involved; how the ancients used electricity; megalithic building techniques; the use of crystal lenses and the fire from the gods; evidence of various high tech weapons in the past, including atomic weapons; ancient metallurgy and heavy machinery; the role of modern inventors such as Nikola Tesla in bringing ancient technology back into modern use; impossible artifacts; and more.

356 PAGES. 6X9 PAPERBACK. ILLUSTRATED. BIBLIOGRAPHY. $16.95. CODE: TGOD

Vimana Aircraft of Ancient India & Atlantis

VIMANA AIRCRAFT OF ANCIENT INDIA & ATLANTIS

by David Hatcher Childress, introduction by Ivan T. Sanderson

Did the ancients have the technology of flight? In this incredible volume on ancient India, authentic Indian texts such as the *Ramayana* and the *Mahabharata* are used to prove that ancient aircraft were in use more than four thousand years ago. Included in this book is the entire Fourth Century BC manuscript *Vimaanika Shastra* by the ancient author Maharishi Bharadwaaja, translated into English by the Mysore Sanskrit professor G.R. Josyer. Also included are chapters on Atlantean technology, the incredible Rama Empire of India and the devastating wars that destroyed it. Also an entire chapter on mercury vortex propulsion and mercury gyros, the power source described in the ancient Indian texts. Not to be missed by those interested in ancient civilizations or the UFO enigma.

334 PAGES. 6X9 PAPERBACK. ILLUSTRATED. $15.95. CODE: VAA

LOST CONTINENTS & THE HOLLOW EARTH

I Remember Lemuria and the Shaver Mystery

by David Hatcher Childress & Richard Shaver

Lost Continents & the Hollow Earth is Childress' thorough examination of the early hollow earth stories of Richard Shaver and the fascination that fringe fantasy subjects such as lost continents and the hollow earth have had for the American public. Shaver's rare 1948 book *I Remember Lemuria* is reprinted in its entirety, and the book is packed with illustrations from Ray Palmer's *Amazing Stories* magazine of the 1940s. Palmer and Shaver told of tunnels running through the earth—tunnels inhabited by the Deros and Teros, humanoids from an ancient spacefaring race that had inhabited the earth, eventually going underground, hundreds of thousands of years ago. Childress discusses the famous hollow earth books and delves deep into whatever reality may be behind the stories of tunnels in the earth. Operation High Jump to Antarctica in 1947 and Admiral Byrd's bizarre statements, tunnel systems in South America and Tibet, the underground world of Agartha, the belief of UFOs coming from the South Pole, more.

344 PAGES. 6X9 PAPERBACK. ILLUSTRATED. $16.95. CODE: LCHE

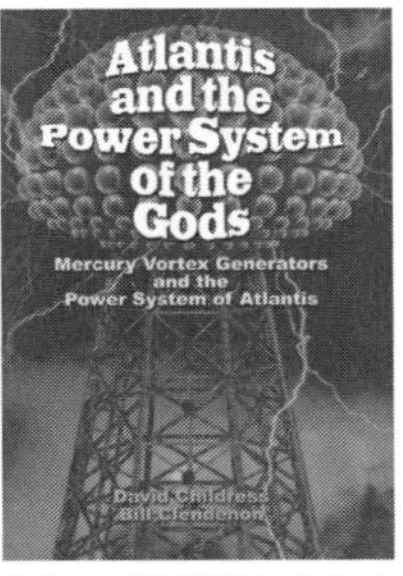

ATLANTIS & THE POWER SYSTEM OF THE GODS

Mercury Vortex Generators & the Power System of Atlantis

by David Hatcher Childress and Bill Clendenon

Atlantis and the Power System of the Gods starts with a reprinting of the rare 1990 book *Mercury: UFO Messenger of the Gods* by Bill Clendenon. Clendenon takes on an unusual voyage into the world of ancient flying vehicles, strange personal UFO sightings, a meeting with a "Man In Black" and then to a centuries-old library in India where he got his ideas for the diagrams of mercury vortex engines. The second part of the book is Childress' fascinating analysis of Nikola Tesla's broadcast system in light of Edgar Cayce's "Terrible Crystal" and the obelisks of ancient Egypt and Ethiopia. Includes: Atlantis and its crystal power towers that broadcast energy; how these incredible power stations may still exist today; inventor Nikola Tesla's nearly identical system of power transmission; Mercury Proton Gyros and mercury vortex propulsion; more. Richly illustrated, and packed with evidence that Atlantis not only existed—it had a world-wide energy system more sophisticated than ours today.

246 PAGES. 6X9 PAPERBACK. ILLUSTRATED. $15.95. CODE: APSG

A HITCHHIKER'S GUIDE TO ARMAGEDDON

by David Hatcher Childress

With wit and humor, popular Lost Cities author David Hatcher Childress takes us around the world and back in his trippy finalé to the Lost Cities series. He's off on an adventure in search of the apocalypse and end times. Childress hits the road from the fortress of Megiddo, the legendary citadel in northern Israel where Armageddon is prophesied to start. Hitchhiking around the world, Childress takes us from one adventure to another, to ancient cities in the deserts and the legends of worlds before our own. Childress muses on the rise and fall of civilizations, and the forces that have shaped mankind over the millennia, including wars, invasions and cataclysms. He discusses the ancient Armageddons of the past, and chronicles recent Middle East developments and their ominous undertones. In the meantime, he becomes a cargo cult god on a remote island off New Guinea, gets dragged into the Kennedy Assassination by one of the "conspirators," investigates a strange power operating out of the Altai Mountains of Mongolia, and discovers how the Knights Templar and their off-shoots have driven the world toward an epic battle centered around Jerusalem and the Middle East.

320 PAGES. 6X9 PAPERBACK. ILLUSTRATED. BIBLIOGRAPHY. INDEX. $16.95. CODE: HGA

One Adventure Place
P.O. Box 74
Kempton, Illinois 60946
United States of America
Tel.: 815-253-6390 • Fax: 815-253-6300
Email: auphq@frontiernet.net
http://www.adventuresunlimitedpress.com
or www.adventuresunlimited.nl

ORDERING INSTRUCTIONS

Remit by USD$ Check, Money Order or Credit Card
Visa, Master Card, Discover & AmEx Accepted
Prices May Change Without Notice
10% Discount for 3 or more Items

SHIPPING CHARGES

United States

Postal Book Rate { $3.00 First Item; 50¢ Each Additional Item
Priority Mail { $4.50 First Item; $2.00 Each Additional Item
UPS { $5.00 First Item; $1.50 Each Additional Item
NOTE: UPS Delivery Available to Mainland USA Only

Canada

Postal Book Rate { $6.00 First Item; $2.00 Each Additional Item
Postal Air Mail { $8.00 First Item; $2.50 Each Additional Item
Personal Checks or Bank Drafts MUST BE USD$ and Drawn on a US Bank
Canadian Postal Money Orders OK
Payment MUST BE USD$

All Other Countries

Surface Delivery { $10.00 First Item; $4.00 Each Additional Item
Postal Air Mail { $14.00 First Item; $5.00 Each Additional Item
Payment MUST BE USD$
Checks and Money Orders MUST BE USD$ and Drawn on a US Bank or branch.
Add $5.00 for Air Mail Subscription to Future *Adventures Unlimited* Catalogs

SPECIAL NOTES

RETAILERS: Standard Discounts Available
BACKORDERS: We Backorder all Out-of-Stock Items Unless Otherwise Requested
PRO FORMA INVOICES: Available on Request
VIDEOS: NTSC Mode Only. Replacement only.
For PAL mode videos contact our other offices:

Please check: ☑

❒ This is my first order ❒ I have ordered before

Name
Address
City
State/Province Postal Code
Country
Phone day Evening
Fax

Item Code	Item Description	Qty	Total

Please check: ☑

Subtotal ➡
Less Discount-10% for 3 or more items ➡
❒ Postal-Surface — Balance ➡
❒ Postal-Air Mail (Priority in USA) — Illinois Residents 6.25% Sales Tax ➡
Previous Credit ➡
❒ UPS (Mainland USA only) — Shipping ➡
Total (check/MO in USD$ only) ➡

❒ Visa/MasterCard/Discover/Amex

Card Number
Expiration Date

10% Discount When You Order 3 or More Items!